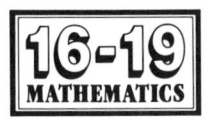

Functions
Unit Guide

The School Mathematics Project

CAMBRIDGE
UNIVERSITY PRESS

Main authors	Simon Baxter
	Stan Dolan
	Doug French
	Andy Hall
	Barrie Hunt
	Lorna Lyons
	Paul Roder
Team leader	Barrie Hunt
Project director	Stan Dolan

The authors would like to give special thanks to Ann White for her help in producing the trial edition and in preparing this book for publication.

Published by the Press Syndicate of the University of Cambridge
The Pitt Building, Trumpington Street, Cambridge CB2 1RP
40 West 20th Street, New York, NY 10011–4211, USA
10 Stamford Road, Oakleigh, Victoria 3166, Australia

© Cambridge University Press 1991

First published 1991
Reprinted 1993

Produced by Gecko Limited, Bicester, Oxon.

Cover design by Iguana Creative Design

Printed in Great Britain at the University Press, Cambridge

British Library cataloguing in publication data

A catalogue record for this book is available from the British Library.

ISBN 0 521 40877 6

Contents

Introduction to 16–19 Mathematics (iv)

Why 16–19 Mathematics?
Structure of the courses
Material

Introduction to the unit (vii)

1 Algebra of functions 1

   Discussion point commentaries
   Tasksheet commentaries

2 Circular functions 15

   Discussion point commentaries
   Tasksheet commentaries

3 Growth functions 27

   Discussion point commentaries
   Tasksheet commentaries

4 Radians 39

   Discussion point commentaries
   Tasksheet commentaries

5 e 45

   Discussion point commentaries
   Tasksheet commentaries

6 Transformations 52

   Discussion point commentaries
   Tasksheet commentaries

Introduction to 16–19 Mathematics

Nobody reads introductions and nobody reads teachers' guides, so what chance does the introduction to this Unit Guide have? The least we can do is to keep it short! We hope that you will find the discussion point and tasksheet commentaries and ideas on presentation and enrichment useful.

The School Mathematics Project was founded in 1961 with the purpose of improving the teaching of mathematics in schools by the provision of new course materials. SMP authors are experienced teachers and each new venture is tested by schools in a draft version before publication. Work on *16–19 Mathematics* started in 1986 and the pilot of the course has been used by over 30 schools since 1987.

Since its inception the SMP has always offered an 'after sales service' for teachers using its materials. If you have any comments on *16–19 Mathematics*, or would like advice on its use, please write to:

> 16–19 Mathematics
> The SMP Office
> The University
> Southampton SO9 5NH

Why 16–19 Mathematics?

A major problem in mathematics education is how to enable ordinary people to comprehend in a few years concepts which geniuses have taken centuries to develop. In theory, our view of how to pass on this body of knowledge effectively and pleasurably has changed considerably; but no great revolution in practice has been seen in sixth-form classrooms generally. We hope that, in this course, the change in approach to mathematics teaching embodied in GCSE schemes will be carried forward. The principles applied in the course are appropriate to this aim.

- Students are actively involved in developing mathematical ideas.
- Premature abstraction and over-reliance on algorithms are avoided.
- Wherever possible, problems arise from, or at least relate to, everyday life.
- Appropriate use is made of modern technology such as graphic calculators and microcomputers.
- Misunderstandings are confronted and acted upon.

By applying these principles and presenting material in an attractive way, A level mathematics is made more accessible to students and more meaningful to them as individuals. The *16–19 Mathematics* course is flexible enough to provide for the whole range of students who obtain at least a grade C at GCSE.

INTRODUCTION TO 16-19 MATHEMATICS

Structure of the courses

The A and AS level courses have a core-plus-options structure. Details of the full range of possibilities, including A and AS level *Further Mathematics* courses, may be obtained from the Joint Matriculation Board, Manchester M15 6EU.

For the A level course *Mathematics (Pure with Applications)*, students must study eight core units and a further two optional units. The structure diagram below shows how the units are related to each other. Other optional units are being developed to give students an opportunity to study aspects of mathematics which are appropriate to their personal interests and enthusiasms.

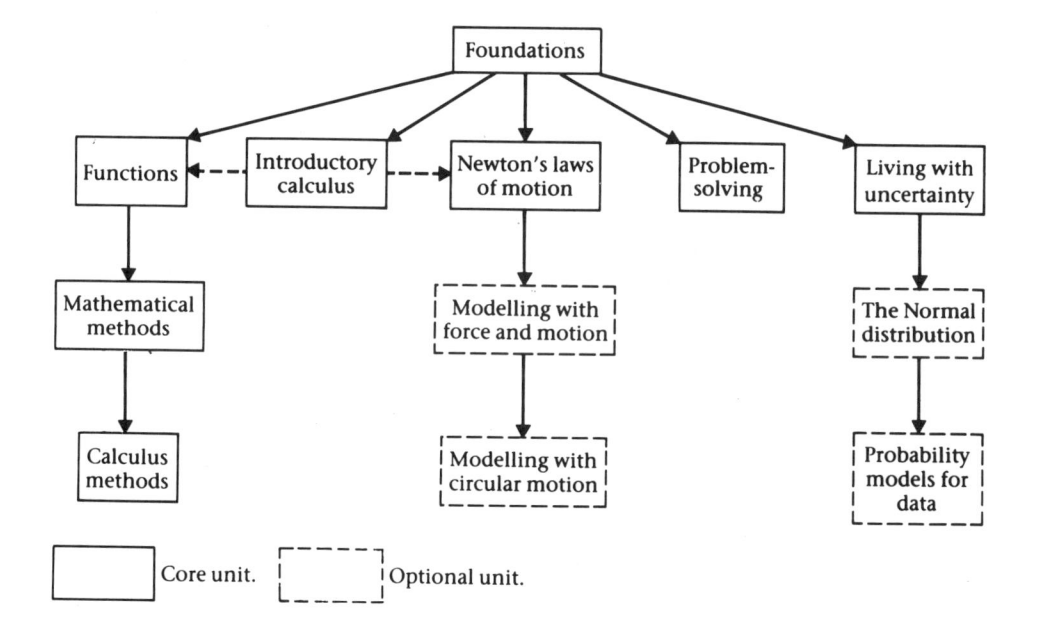

```
                        ┌─────────────┐
                        │ Foundations │
                        └──────┬──────┘
         ┌──────────┬─────────┼─────────┬──────────────┐
         ▼          ▼         ▼         ▼              ▼
   ┌─────────┐ ┌──────────┐ ┌────────┐ ┌────────┐ ┌──────────┐
   │Functions│◄┤Introductory├┤Newton's│ │Problem-│ │Living with│
   │         │ │ calculus │ │laws of │ │solving │ │uncertainty│
   │         │ │          │ │ motion │ │        │ │          │
   └────┬────┘ └──────────┘ └───┬────┘ └────────┘ └────┬─────┘
        ▼                       ▼                      ▼
  ┌──────────────┐       ┌──────────────┐       ┌──────────────┐
  │ Mathematical │       │ Modelling with│       │ The Normal   │
  │   methods    │       │force and motion│      │ distribution │
  └──────┬───────┘       └───────┬──────┘       └──────┬───────┘
         ▼                       ▼                      ▼
  ┌──────────────┐       ┌──────────────┐       ┌──────────────┐
  │  Calculus    │       │ Modelling with│       │ Probability  │
  │  methods     │       │circular motion│       │  models for  │
  │              │       │              │       │    data      │
  └──────────────┘       └──────────────┘       └──────────────┘
```

☐ Core unit.   ⌐ ¬ Optional unit.

──▶ The *Foundations* unit should be started before or at the same time as any other core unit.

- - ▶ Any of the other units can be started at the same time as the *Foundations* unit. The second half of *Functions* requires prior coverage of *Introductory calculus*. *Newton's laws of motion* requires calculus notation which is covered in the initial chapters of *Introductory calculus*.

For the AS level *Mathematics (Pure with Applications)* course, students must study *Foundations, Introductory calculus* and *Functions*. Students must then study a further two applied units.

v

Material

The textbooks contain several new devices to aid an active style of learning.

- Topics are opened up through **group discussion points**, signalled in the text by the symbol

  and enclosed in rectangular frames. These consist of pertinent questions to be discussed by students, with guidance and help from the teacher. Commentaries for discussion points are included in this unit guide.

- The text is also punctuated by **thinking points**, having the shape

  and again containing questions. These should be dealt with by students without the aid of the teacher. In facing up to the challenge offered by the thinking points it is intended that students will achieve a deeper insight and understanding. A solution within the text confirms or modifies the student's response to each thinking point.

- At appropriate points in the text, students are referred to **tasksheets** which are placed at the end of the relevant chapter. A tasksheet usually consists of a self-contained piece of work which is used to investigate a concept prior to any formal exposition. In many cases, it takes up an idea raised in a discussion point, examining it in more detail and preparing the way for formal treatment. There are also **extension tasksheets** (labelled by an E), for higher attaining students, which investigate a topic in more depth and **supplementary tasksheets** (labelled by an S), which are intended to help students with a relatively weak background in a particular topic. Commentaries for all the tasksheets are included in this unit guide.

The aim of the **exercises** is to check full understanding of principles and give the student confidence through reinforcement of his or her understanding.

Graphic calculators/microcomputers are used throughout the course. In particular, much use is made of graph plotters. The use of videos and equipment for practical work is also recommended.

As well as the textbooks and unit guides, there is a *Teacher's resource file*. This file contains: review sheets which may be used for homework or tests; datasheets; technology datasheets which give help with using particular calculators or pieces of software; a programme of worksheets for more able students which would, in particular, help prepare them for the STEP examination.

Introduction to the unit

Following an introductory chapter, the central theme of the unit is the concept of a function. The properties of trigonometric, exponential and logarithmic functions and their derived functions are investigated. An important aim of the unit is to give students a clear understanding of why e is a 'natural' base for exponential functions and why radians are a 'natural' measure of angle.

In *Foundations*, students will have met the idea of transforming the graph of a basic function to fit a more complex function. This is built on and consolidated throughout the course and the final chapter pulls together the ideas on transformation met in the previous five chapters, giving the unit coherence.

Using this unit

A graphical approach is adopted wherever appropriate and students are frequently required to use a graph plotter to carry out investigations. Although any graph plotter may be used, the one included in the 16–19 software package is particularly suitable as it enables students to work creatively using function notation. Whichever graph plotter is used, it is essential that it has a facility for plotting the gradient graph of a function. Several results which are investigated by students are not proved rigorously in this unit. A more formal treatment occurs, where necessary, in later pure mathematics units.

Some additional notes on the individual chapters may prove helpful.

Chapter 1

The concepts of function, domain, range, inverse function and composite function are introduced in this chapter. Students also look at translations and reflections of graphs in the context of odd, even and inverse functions. The chapter builds on ideas met previously in *Foundations* and students should be able to proceed at a fairly rapid pace.

Chapter 2

The idea of transforming the graph of $y = \sin x°$ into that of $y = a \sin (bx + c) + d$ by applying a sequence of stretches and translations is investigated and the importance of circular functions as models for natural periodic phenomena is stressed.

Chapter 3

The importance of the exponential function as a model for natural growth and decay is stressed. The log function is introduced as an inverse function for exponential growth.

Chapter 4

Before tackling this chapter and the following one, students must have a good grasp of calculus and should have completed the *Introductory calculus* unit. The approach is investigative and students are expected to have access to a graph plotter with a gradient graph facility. It is emphasised that the results

$$\frac{d}{dx}(\sin x) = \cos x, \qquad \frac{d}{dx}(\cos x) = -\sin x$$

only hold when x is measured in radians.

Chapter 5

The reason for the choice of e as base for exponential functions is investigated, as is the fact that **all** exponential functions may be expressed in the form Ae^{bx}. The complementary property that all logarithmic functions can be expressed using natural logs should be emphasised.

Chapter 6

The short final chapter should take no more than a couple of lessons. Stretches, reflections and translations of a graph form a theme connecting the previous five chapters. This chapter deals with transformations applied to a circle and pulls together ideas met throughout the unit.

Tasksheets and resources

Items in *italics* refer to resources not included in the main text.

1 Algebra of functions

1.1 Composition of functions
 Tasksheet 1S – Functions of functions
 Tasksheet 2E – Functions of functions
1.2 Range and domain
1.3 Inverse functions
 Tasksheet 3 – Inverse functions
1.4 Rearranging formulas
 Tasksheet 4S – Rearranging formulas
1.5 Parameters and functions
 Tasksheet 5 – Parametric formulas
1.6 Functions and transformations of graphs
 Tasksheet 6 – Translations of graphs
1.7 Combining transformations of graphs
 Tasksheet 7 – Combining transformations

2 Circular functions

2.1 Rotation
 Tasksheet 1S – Sin $\theta°$ and cos $\theta°$
 Datasheet 1 – Sin $\theta°$ and cos $\theta°$
2.2 Transformations
 Tasksheet 2 – Transformations
2.3 Modelling periodic behaviour
 Tasksheet 3 – Applications
2.4 Inverse trigonometric functions
 Tasksheet 4E – Inverse functions
2.5 Solving equations
 Tasksheet 5S – Solving equations
2.6 Tan $\theta°$
 Tasksheet 6S – Tan $\theta°$

3 Growth functions

3.1 Exponential growth
3.2 Indices
 Tasksheet 1 – Indices
 Tasksheet 2S – Laws of indices
3.3 Growth factors
 Tasksheet 3 – Ka^x
3.4 Logarithms
 Tasksheet 4 – Properties of logs
3.5 The equation $a^x = b$
 Tasksheet 5 – $a^x = b$

4 Radians

4.1 Rates of change
Tasksheet 1 – Differentiation of sin $\theta°$
4.2 Radian measure
Tasksheet 2 – Radian measure
4.3 Area and arc lengths
4.4 More about derivatives
Tasksheet 3 – Further derivatives
Tasksheet 4E – Derivative of $\sin^2 x$
4.5 Applications

5 e

5.1 e^x
Tasksheet 1 or 1E – Rates of growth
5.2 e^{ax}
Tasksheet 2 – e^{ax}
5.3 The natural log
Tasksheet 3 – Properties of ln x

6 Transformations

6.1 Graph sketching
Tasksheet 1 – Transforming equations

6.2 Stretching a circle

1 Algebra of functions

1.1 Composition of functions

(a) What is (i) 122°F in °C, (ii) 300°C in K, (iii) 122°F in K?
(b) What is the general rule for converting °F into K?

(a) (i) 50°C (ii) 573 K (iii) 323 K

(b) The output in °C is used as input to g(t) in place of t. Therefore the rule is $t \to \frac{5}{9}(t-32) + 273$.

1.3 Inverse functions

(a) If $f(x) = x^2$, what are $f(-3)$ and $f(+3)$?
What is meant by $f^{-1}(9)$? Are there two possible answers?
(b) What is meant by $\sqrt{x}$?
Does the same problem arise with $\sqrt[3]{x}$?

(a) $f(-3) = f(+3) = 9$

$f^{-1}(9)$ is ambiguous, since it could mean both -3 and $+3$. A function must be unambiguous – a single input must give rise to a **unique** output. Thus f^{-1} is not a function if the domain of f is unrestricted.

The inverse function does exist if the domain is restricted to non-negative numbers.

In general, a function must be one-to-one in order to have an inverse.

(b) $\sqrt{x}$ means the non-negative number whose square is x.

The same problem does not arise with $\sqrt[3]{x}$ because $x \to x^3$ is a one-to-one function.

1.6 Functions and transformations of graphs

(a) Which of these graphs can be mapped onto other graphs in the diagram?

(b) What transformations would map

  (i) graph b onto graph c,

  (ii) graph a onto graph d,

  (iii) graph b onto graph e?

(c) Is there more than one possible answer to any of questions (i) to (iii)?

(a) Graphs b, c and e can be mapped onto each other by translations. Graphs a and d can be mapped onto each other by translations.

(b) (i) The translation through $\begin{bmatrix} 5\frac{1}{2} \\ -2\frac{3}{4} \end{bmatrix}$

  (ii) The translation through $\begin{bmatrix} 7 \\ -2 \end{bmatrix}$

  (iii) The translation through $\begin{bmatrix} 11 \\ 0 \end{bmatrix}$

(c) b can be mapped onto e by a reflection in the y-axis.

1.7 Combining transformations of graphs

(a) Which of these graphs can be mapped onto other graphs in the diagram?

(b) What transformations would map

(i) graph *b* onto graph *c*,

(ii) graph *a* onto graph *d*,

(iii) graph *c* onto graph *e*,

(iv) graph *b* onto graph *e*?

(c) Is there more than one possible answer to any of questions (i) to (iv)?

(a) Graphs *b*, *c* and *e* can be mapped onto each other by translations, reflections or rotations. Graphs *a* and *d* can be mapped onto each other by a reflection or a half-turn.

(b),(c) (i) A translation by $\begin{bmatrix} 11 \\ 0 \end{bmatrix}$ or a reflection in the *y*-axis

(ii) Reflection in the *x*-axis or half-turn about the origin

(iii) A half-turn about the point $(5\frac{1}{2}, 0)$ or a reflection in the *x*-axis

(iv) A half-turn about the origin

Functions of functions

TASKSHEET COMMENTARY 1S

1 (a) (i) $f(4) = 4^2 = 16$ (ii) $g(16) = 3 \times 16 + 1 = 49$

(b) $g(f(4)) = g(16) = 49$

(c) (i) $g(f(2)) = g(4) = 13$ (ii) $g(f(-3)) = g(9) = 28$

(iii) $f(g(-2)) = f(-5) = 25$

(d) $x \longrightarrow \boxed{f} \xrightarrow{x^2} \boxed{g} \xrightarrow{3(x^2) + 1 = g(f(x))}$

(e) For example, from (a) $g(f(2)) = 13$, and $3(2^2) + 1 = 13$

(f) $x \longrightarrow \boxed{g} \xrightarrow{3x+1} \boxed{f} \xrightarrow{(3x+1)^2} f(g(x)) = (3x+1)^2$

2 (a) (i) $gf(x) = \dfrac{1}{x} - 3$, $fg(x) = \dfrac{1}{x-3}$

(ii) $gf(x) = \sqrt{(2x)}$, $fg(x) = 2\sqrt{x}$

(iii) $gf(x) = (x+5) - 9 = x - 4$, $fg(x) = (x-9) + 5 = x - 4$

(iv) $gf(x) = fg(x) = 10 - (10 - x) = x$

(v) $gf(x) = fg(x) = \dfrac{1}{1/x} = x$

(b) When $fg(x) = gf(x)$, the order in which the functions are applied does not matter.

In (iii), the order in which successive addition and subtraction is carried out does not matter.

In (iv) and (v), the composite function is in both cases the identity function. Since functions f and g are the same in each example, they are both **self-inverse** functions. (See section 1.3 and tasksheet 3.)

3 (a) $x + 2$ (b) x^2 (c) $\dfrac{1}{x}$ (d) $\sqrt{x}$

4 (a) $x + 8$ (b) x^2 (c) $3x + 1$ (d) $\dfrac{12}{x}$ (e) $\sqrt[3]{x}$ (f) $4x - x^2$

Functions of functions

TASKSHEET COMMENTARY 2E

1 (a) $x + 4$ (b) x^4 (c) $4x - 9$
(d) x (e) $\sin(\sin x)$ (f) x

2 (a) $f(g(f(x))) = f(g(x - 3)) = f((x - 3)^2) = (x - 3)^2 - 3$
(b) (i) $fg(x)$ (ii) $gfg(x)$ (iii) $f^2(x)$ (iv) $fg^3(x)$ (v) $f^2g^2f(x)$

3 (a) $qs(x)$ (b) $sq(x)$ (c) $s^2(x)$

4 (a) $fg(x) = (x + 3)^2$, $gf(x) = x^2 + 3$, $x = -1$
(b) $fg(x) = x - 3$, $gf(x) = x - 3$, all values of x
(c) $fg(x) = 6x + 1$, $gf(x) = 6x - 2$, no values of x
(d) $fg(x) = \dfrac{1}{x^3}$, $gf(x) = \dfrac{1}{x^3}$, all values of x
(e) $fg(x) = x + 2$, $gf(x) = x + 1$, no values of x
(f) $fg(x) = \sqrt{(x - 1)}$, $gf(x) = \sqrt{(x)} - 1$, $x = 1$

5

	e	f	g	h
e	e	f	g	h
f	f	e	h	g
g	g	h	e	f
h	h	g	f	e

1 ALGEBRA OF FUNCTIONS

Inverse functions

TASKSHEET COMMENTARY 3

1 (a) f^{-1} reverses the effect of f and so $f^{-1}(f(x)) = x$.

(b) If $f(x) = y$, then $f^{-1}(y) = x$. Any point in the domain of f is therefore in the range of f^{-1} and vice versa.

(c)

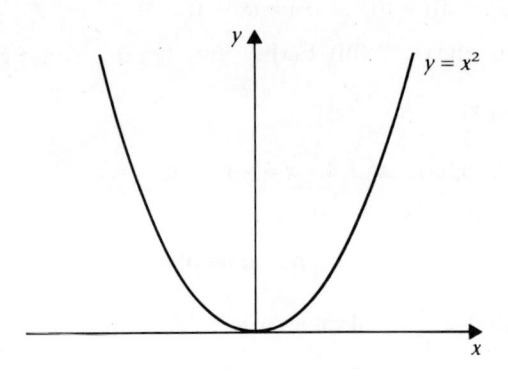

[Graph showing $y = x^2$ parabola]

(d) If a function is many-to-one (as is $y = x^2$ in (c)), then the inverse is not a function because it is one-to-many.

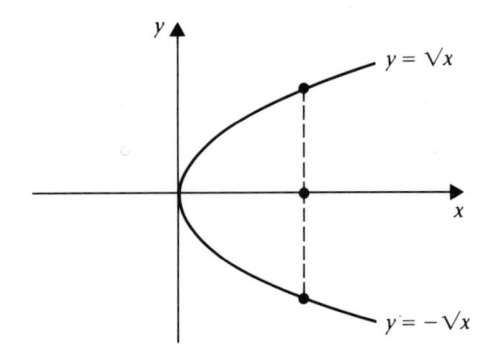

[Graph showing sideways parabola with $y = \sqrt{x}$ and $y = -\sqrt{x}$]

2 (a)

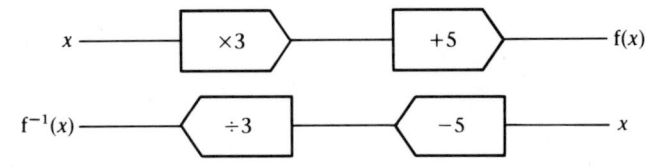

$x \longrightarrow [\times 3] \longrightarrow [+5] \longrightarrow f(x)$

$f^{-1}(x) \longrightarrow [\div 3] \longrightarrow [-5] \longrightarrow x$

$f^{-1}(x) = \frac{1}{3}(x - 5)$, e.g. $f^{-1}(f(2)) = f^{-1}(11) = 2$ as required.

(b) $g^{-1}(x) = \sqrt{(x + 7)}$ (c) $h^{-1}(x) = \sqrt{x} + 7$ (d) $r^{-1}(x) = (x - 6)^2$

TASKSHEET COMMENTARY 3

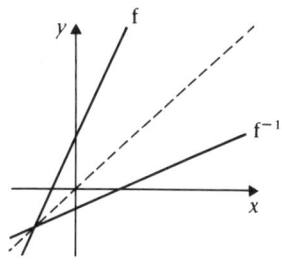

3 (i) (a) Domain of f = $\mathbb{R}$

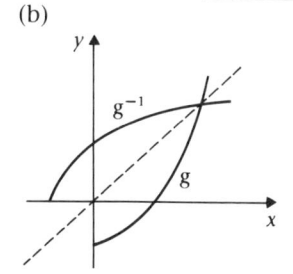

(b) Domain of g = $\{x \in \mathbb{R} : x \geq 0\}$

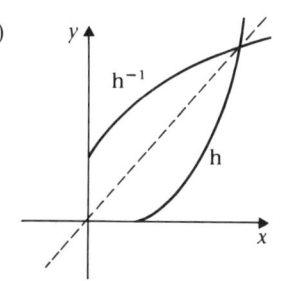

(c) Domain of h = $\{x \in \mathbb{R} : x \geq 7\}$

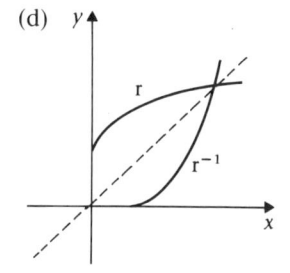

(d) Domain of r = $\{x \in \mathbb{R} : x \geq 0\}$

(ii) Reflection in $y = x$ (resulting in interchange of x and y in the equation).

4 (a) $f^{-1}(x) = \sqrt{(x+3)} - 5$ (b) $f^{-1}(x) = \frac{1}{2}(\frac{1}{3}x + 1)$

5 (a) (i) The reciprocal sequence alternates between two values, except when $x_1 = 0$ (x_2 undefined) or $x_1 = \pm 1$ (constant sequence).

(ii) The 'change sign' sequence behaves similarly; it is constant for $x_1 = 0$.

(b) Both functions are self-inverse and hence $f^2(x) = x$ in each case.

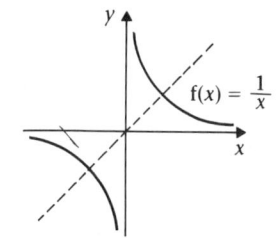

$f(x) = \frac{1}{x}$ $f(x) = -x$

The graphs of both functions reflect in the line $y = x$ onto themselves.

6 (a) $\dfrac{1-x}{2x}$ (b) $12 - x$ (c) $\dfrac{1}{x+1}$ (d) $\sqrt{\left(\dfrac{8}{x}\right) - 1}$

(e) $\sqrt{(1-x^2)}$ (f) $\sqrt{(4-x)} + 2$

(b) and (e) are self-inverse and have $y = x$ as a line of symmetry.

Rearranging formulas

TASKSHEET COMMENTARY 4S

1
$$\frac{y}{5} = (x-7)^2$$

$$\pm\sqrt{\left(\frac{y}{5}\right)} = (x-7)$$

$$\pm\sqrt{\left(\frac{y}{5}\right)} + 7 = x$$

2 (a) $\quad y + 7 = 3x^2$

$$\Rightarrow \frac{1}{3}(y+7) = x^2 \qquad \left[\text{or } \frac{y+7}{3} = x^2\right]$$

$$\Rightarrow \quad x = \pm\sqrt{\left(\frac{1}{3}(y+7)\right)} \qquad \left[\text{or } \pm\sqrt{\left(\frac{y+7}{3}\right)}\right]$$

(b) $9y = (2x+1)^2$

$\Rightarrow \pm\sqrt{(9y)} = 2x + 1 \quad$ [and $\sqrt{(9y)} = \sqrt{9}\sqrt{y} = 3\sqrt{y}$]

$\Rightarrow \pm\sqrt{(9y)} - 1 = 2x$

$\Rightarrow x = \frac{1}{2}(\pm\sqrt{(9y)} - 1) \qquad \left[\text{or } \frac{1}{2}(\pm 3\sqrt{y} - 1)\text{, from above}\right]$

(c) $y + 1 = 3\sqrt{x}$

$$\Rightarrow \frac{1}{3}(y+1) = \sqrt{x}$$

$$\Rightarrow x = \frac{1}{9}(y+1)^2$$

3
$$yx = 3 - 4x$$

$\Rightarrow yx + 4x = 3$

$\Rightarrow (y+4)x = 3$

$$\Rightarrow \quad x = \frac{3}{y+4}$$

TASKSHEET COMMENTARY 4S

NB. an alternative approach to problems involving reciprocals is:

$$y + 4 = \frac{3}{x} \quad \text{[adding 4 to both sides]}$$

$$\Rightarrow \frac{y+4}{3} = \frac{1}{x} \quad \left[\text{dividing both sides by 3, since } \frac{3}{x} = 3 \times \left(\frac{1}{x}\right)\right]$$

$$\Rightarrow \frac{3}{y+4} = \frac{x}{1} = x \quad \text{[inverting both sides]}$$

4 (a) $\quad xy = 2x - 1$

$$\Rightarrow xy - 2x = -1$$

$$\Rightarrow x(y - 2) = -1$$

$$\Rightarrow x = \frac{-1}{y-2}$$

$$\Rightarrow x = \frac{1}{2-y} \quad \text{[multiplying top and bottom by } -1\text{]}$$

(b) $\quad yx^2 = 3$

$$\Rightarrow x^2 = \frac{3}{y}$$

$$\Rightarrow x = \pm\sqrt{\left(\frac{3}{y}\right)}$$

(c) $\quad 2y\sqrt{x} = 10\sqrt{x} + 1$

$$\Rightarrow 2y\sqrt{x} - 10\sqrt{x} = 1$$

$$\Rightarrow 2\sqrt{x}(y - 5) = 1$$

$$\Rightarrow \sqrt{x} = \frac{1}{2(y-5)}$$

$$\Rightarrow x = \frac{1}{4(y-5)^2}$$

Parametric formulas

TASKSHEET COMMENTARY 5

1 $\qquad l = l_0(1 + \alpha t)$

$\Rightarrow \quad \dfrac{l}{l_0} = 1 + \alpha t$

$\Rightarrow \quad \dfrac{l}{l_0} - 1 = \alpha t$

$\Rightarrow \quad \dfrac{1}{t}\left(\dfrac{l}{l_0} - 1\right) = \alpha$

$\left[\text{more significantly, } \alpha = \dfrac{1}{t}\left(\dfrac{l - l_0}{l_0}\right); (l - l_0) \text{ being the increase in length}\right]$

$\qquad \alpha = \dfrac{1}{230}\left(\dfrac{1.004}{1} - 1\right) = 1.7 \times 10^{-5}$

2 $\dfrac{E}{m} = c^2 \Rightarrow c = \sqrt{\left(\dfrac{E}{m}\right)}$ [since c is known to be positive]

3 $\quad E + \tfrac{1}{2}mu^2 = \tfrac{1}{2}mv^2$

$\Rightarrow 2E + mu^2 = mv^2$

$\Rightarrow \dfrac{2E + mu^2}{m} = v^2 \qquad \left[\text{or } v^2 = \dfrac{2E}{m} + u^2\right]$

$\Rightarrow \qquad v = \sqrt{\left(\dfrac{2E + mu^2}{m}\right)} \qquad \left[\text{or } v = \sqrt{\left(\dfrac{2E}{m} + u^2\right)}\right]$

4 $\quad T = 2\pi\sqrt{\left(\dfrac{l}{g}\right)}$

$\Rightarrow \dfrac{T}{2\pi} = \sqrt{\left(\dfrac{l}{g}\right)}$

$\Rightarrow \dfrac{T^2}{4\pi^2} = \dfrac{l}{g}$

$\Rightarrow \dfrac{T^2 g}{4\pi^2} = l$

$l = \dfrac{2^2 \times 9.81}{4\pi^2} = 0.9940$

TASKSHEET COMMENTARY 5

5 (a) Area of cylindrical surface $= 2\pi rh$
Area of ends $= 2\pi r^2$
Total surface area $= 2\pi rh + 2\pi r^2$
$S = 2\pi r(h + r)$

(b) $\dfrac{S}{2\pi r} = h + r \Rightarrow h = \dfrac{S}{2\pi r} - r$

6 (a) $r = \sqrt{\left(\dfrac{3V}{\pi h}\right)}$ (b) $r = \dfrac{100I}{Pn}$

7 (a) $I = \dfrac{nE}{R + nr} \Rightarrow I(R + nr) = nE$

$\Rightarrow IR = nE - nrI$

$\Rightarrow R = \dfrac{n(E - rI)}{I}$

(b) $E = \tfrac{1}{2}mv^2 \Rightarrow 2E = mv^2$

$\Rightarrow \dfrac{2E}{m} = v^2$

$\Rightarrow v = \sqrt{\left(\dfrac{2E}{m}\right)}$

Translations of graphs

TASKSHEET COMMENTARY 6

1 (a) $f(x) = x^4$, $f(x) + 2 = x^4 + 2$, $f(x+3) = (x+3)^4$

(b) The three graphs are congruent and the transformations needed are

(i) translation $\begin{bmatrix} 0 \\ 2 \end{bmatrix}$, (ii) translation $\begin{bmatrix} -3 \\ 0 \end{bmatrix}$.

2 $g(x) = \dfrac{1}{x}$, $g(x+4) = \dfrac{1}{x+4}$, $g(x+4) + 3 = \dfrac{1}{x+4} + 3$

The three graphs are congruent and the transformations needed are

(i) translation $\begin{bmatrix} -4 \\ 0 \end{bmatrix}$, (ii) translation $\begin{bmatrix} -4 \\ 3 \end{bmatrix}$.

3 In questions 1 and 2 you saw that replacing 'x' by 'x − 5' resulted in a translation of 5 parallel to the x-axis; and that adding 2 to the function was equivalent to a translation of 2 parallel to the y-axis.

So it is reasonable to suggest that the image of $y = \dfrac{3}{x^2}$ under a translation $\begin{bmatrix} 5 \\ 2 \end{bmatrix}$ is

$$y = \dfrac{3}{(x-5)^2} + 2$$

4 $y = (x+2)^2 - 1$, a translation by $\begin{bmatrix} -2 \\ -1 \end{bmatrix}$

1 ALGEBRA OF FUNCTIONS

Combining transformations

TASKSHEET COMMENTARY 7

1 (a) $f(x) = x^2 - x$, $f(-x) = x^2 + x$, $-f(x) = -x^2 + x$

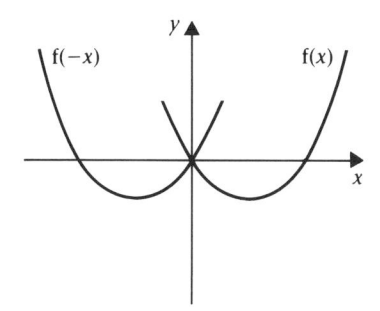

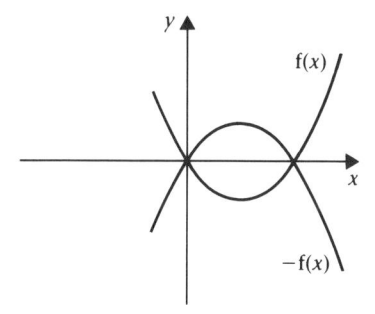

(b) (i) reflection in y-axis   (ii) reflection in x-axis

(c) Yes

   (i) If (a, b) is a point on the graph of $y = f(x)$, then $(-a, b)$ will be a point on the graph of $y = f(-x)$.

   (ii) Similarly, if (a, b) is on the graph of $y = f(x)$, then $(a, -b)$ is on the graph of $y = -f(x)$.

2 (a) The equation of the reflected graph will be $y = -f(x) = -x^4 + 2x^3$.

(b) The equation of the reflected graph will be $y = f(-x) = x^4 + 2x^3$.

13

TASKSHEET COMMENTARY 7

3 (a) $f(-x) = 3x^2 - x^4$

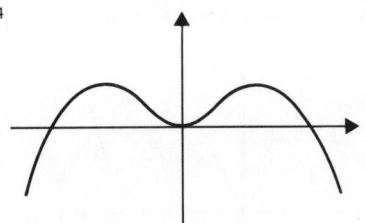

The graphs of $f(x)$ and $f(-x)$ coincide, since $(-x)^2 = x^2$ and $(-x)^4 = x^4$.

(b) $f(-x) = -x^3 + 5x = -f(x)$

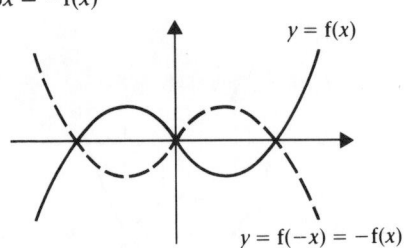

Here $f(-x) = -f(x)$, since $(-x)^3 = -x^3$ and $5(-x) = -5x$.

4 (a) Even (b) odd (c) neither

5 (a) Even (b) neither (c) neither
(d) odd (e) odd (f) even

6 (a) After the first reflection:
$$y = -f(x) = -x^2 - 3x + 2 = g(x)$$
After the second reflection:
$$y = g(-x) = -x^2 + 3x + 2$$

(b) A single equivalent transformation is a 180° rotation about the origin.

7 The equation of the curve obtained by reflection is
$$y = f(-x) = 2x^2 + \frac{1}{x} = g(x)$$

The equation of the curve obtained after translation is
$$y = g(x - 4) + 3 = 2(x - 4)^2 + \frac{1}{x - 4} + 3$$

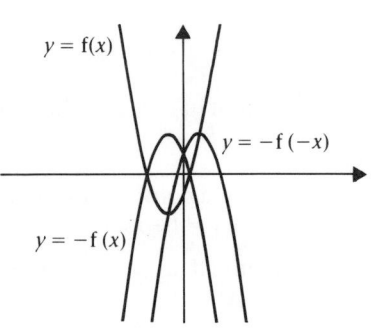

2 Circular functions

2.1 Rotation

(a) Explain why $(\cos \theta°, \sin \theta°)$ may be defined as the coordinates of H and why $h = \sin \theta°$.

(b) Is this definition still appropriate when
   (i) θ is greater than 360,   (ii) θ is negative?

(c) What must the cowboy do with the lasso to obtain a negative angle?

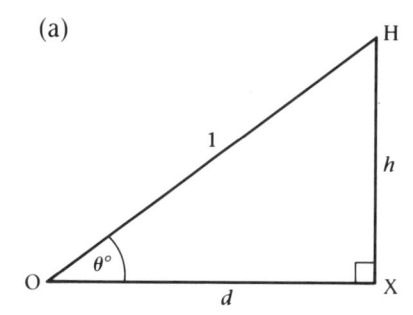

(a) In a right-angled triangle for which the hypotenuse is of length 1,

$$\sin \theta° = \frac{h}{1} \quad \text{and} \quad \cos \theta° = \frac{d}{1}$$

i.e. $\sin \theta° = h$ and $\cos \theta° = d$.

This definition of $\sin \theta°$ and $\cos \theta°$ is, however, restricted to values of θ between 0 and 90. In order to consider values of θ outside this range, a new definition is needed. This definition must be **consistent** with that for 0 to 90, i.e. it must give the same value within that range. If you define $(\cos \theta°, \sin \theta°)$ as the coordinates of H for any angle θ, then the values given by the more restricted definition are retained.

(b) This definition is equally appropriate to angles greater than 360° or to negative angles since the point H is clearly defined in all cases.

(c) If the lasso moves in a clockwise direction it will give a negative angle.

2.2 Transformations

> (a) What are the equations of the four graphs?
>
> (b) What do the questions above tell you about the relationship between
>
>   (i) sin $(-\theta)°$ and sin $\theta°$,   (ii) cos $(-\theta)°$ and cos $\theta°$?
>
> (c) Are the sine and cosine functions odd, even, or neither odd nor even?

(a) graph A: $y = \sin \theta°$;   graph B: $y = \cos \theta°$;   graph C: $y = -\sin \theta°$;   graph D: $y = -\cos \theta°$

(b) (i) $\sin(-\theta)° = -\sin \theta°$ (ii) $\cos(-\theta)° = \cos \theta°$

(c) The sine function is odd, the cosine is even.

2.4 Inverse trigonometric functions

> (a) Use your calculator to find a solution to the equation $\sin x° = 0.6$.
>
> (b) How many more solutions can you find to this equation?
>
> (c) Why is the inverse of f, where $f(x) = \sin x°$ ($x \in \mathbb{R}$), **not** a function?

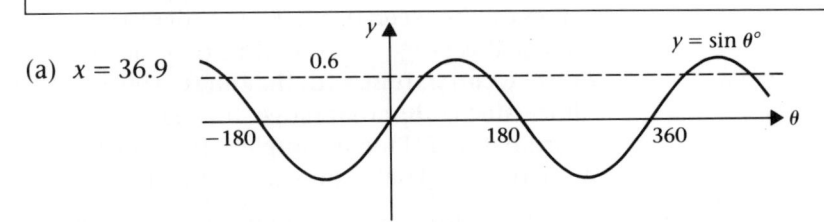

(a) $x = 36.9$

(b) From the graph there is clearly an infinite number of solutions. They include

$180 - 36.9 = 143.1$, $360 + 36.9 = 396.9$, $360 + 143.1 = 503.1$,

$36.9 - 360 = -323.1$, $143.1 - 360 = -216.9$

The **general solution** is $180n + (-1)^n 36.9$. The idea of a general solution is not developed at this stage, apart from on the extension tasksheet 4E.

(c) In general, a mapping from a set A to a set B is a function if and only if every element a of set A has a unique image in set B. For continuous functions of real numbers, this means that a vertical line drawn on the graph must cut the graph at exactly one point.

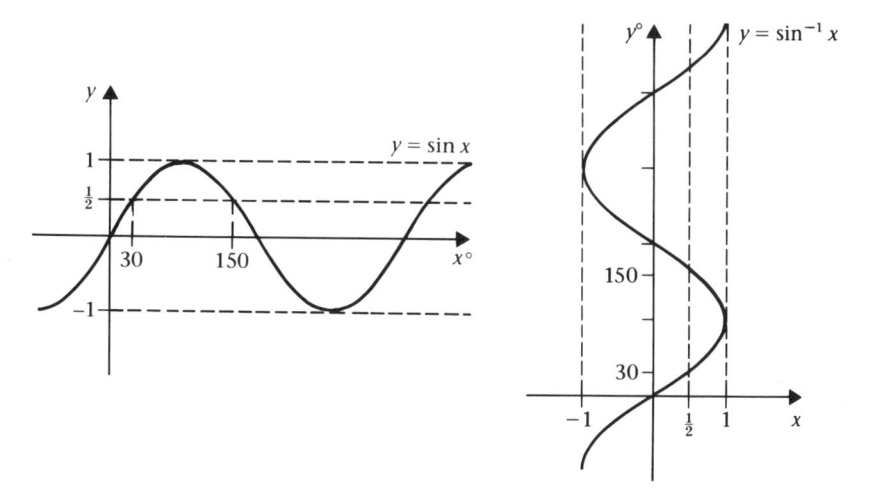

$\sin x$ **is** a function. It is a many-to-one mapping.

From the graph you can see that $\sin^{-1} x$ is **not** a function. For example, the image of $\frac{1}{2}$ under $\sin^{-1}$ includes 30, 150, 390, ... $\sin^{-1}$ is a one-to-many mapping.

The function called $\sin^{-1}$ later in this section is sometimes called arcsin to avoid confusion with the one-to-many mapping.

2.5 Solving equations

(a) When is the height of the tide 6 m?

(b) If a boat can only enter and leave the harbour when the depth of water exceeds 6 m, for how long each day is this possible?

(a) If $h = 6$, $\quad 6 = 2.5 \sin 30t + 5 \Rightarrow \sin 30t = 0.4$

Using a calculator, $\sin x° = 0.4 \Rightarrow x = 23.58°$ and, using the symmetry of the sine graph, $x = 156.42$ is also a solution. So

$\quad\quad 30t = 23.58$ or 156.42
$\Rightarrow \quad t = 0.786$ hours or 5.214 hours

The height is 6 m at 0047 hours and 0513 hours.

17

2 CIRCULAR FUNCTIONS

You could find subsequent times by extending the range of values of x beyond 360. The next two are

$$x = 383.58 \text{ and } 516.42$$
$$\Rightarrow 30t = 383.58 \text{ and } 516.42$$
$$\Rightarrow t = 12.786 \text{ and } 17.214$$

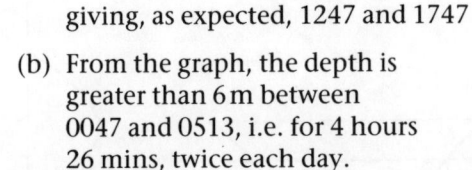

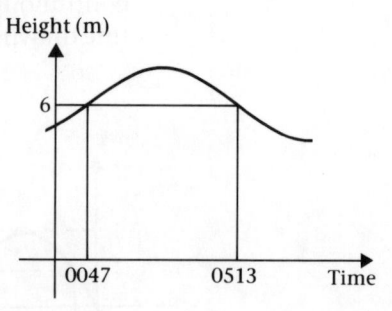

giving, as expected, 1247 and 1747

(b) From the graph, the depth is greater than 6 m between 0047 and 0513, i.e. for 4 hours 26 mins, twice each day.

2.6 Tan $\theta°$

(a) What is (i) length y, (ii) angle CUQ?

(b) Sketch a graph to show how y varies with θ as the ball travels (i) from C to Q, (ii) from P to Q.

(c) Using the sides of a right-angled triangle, show that if $0 \leqslant \theta < 90$ then

$$\tan \theta° = \frac{\sin \theta°}{\cos \theta°}$$

(d) What is the greatest possible domain for tan?

(e) Find a suitable set of principal values for $\tan^{-1} x$.

(a) (i) $5 \tan \theta°$ (ii) the acute angle whose tangent is 3; $\tan^{-1} 3 \approx 71.6°$

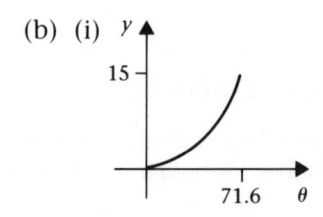

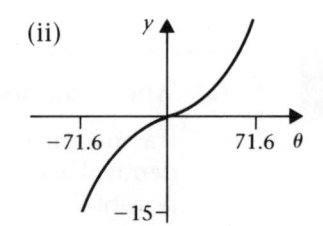

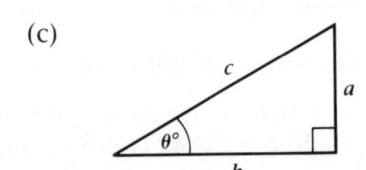

(b) (i) [graph] (ii) [graph]

(c) $\tan \theta° = \dfrac{a}{b}$

$\dfrac{\sin \theta°}{\cos \theta°} = \dfrac{a}{c} \div \dfrac{b}{c} = \dfrac{a}{b}$

(d) All real numbers θ for which $\cos \theta° \neq 0$

(e) $-90° < \tan^{-1} x < 90°$

Sin $\theta°$ and cos $\theta°$

TASKSHEET COMMENTARY 1S

1 (a) $h = \sin 30°$, $(\theta, h) = (30, 0.50)$

   (b) (i) (45, 0.71)   (ii) (60, 0.87)   (iii) (0, 0)   (iv) (90, 1)

2 (a) $h = \sin 120°$, $(\theta, h) = (120, 0.87)$

   (b) (i) (135, 0.71)   (ii) (150, 0.50)   (iii) (180, 0)

3 (a) $h = \sin 210°$, $(\theta, h) = (210, -0.50)$

   (b) (i) (225, −0.71)   (ii) (240, −0.87)   (iii) (270, −1)

4 (a) $h = \sin 300°$, $(\theta, h) = (300, -0.87)$

   (b) (i) (315, −0.71)   (ii) (330, −0.50)   (iii) (360, 0)

5 (390, 0.50), (405, 0.71), (420, 0.87), (450, 1), (480, 0.87). (495, 0.71), (510, 0.50), (540, 0)

6

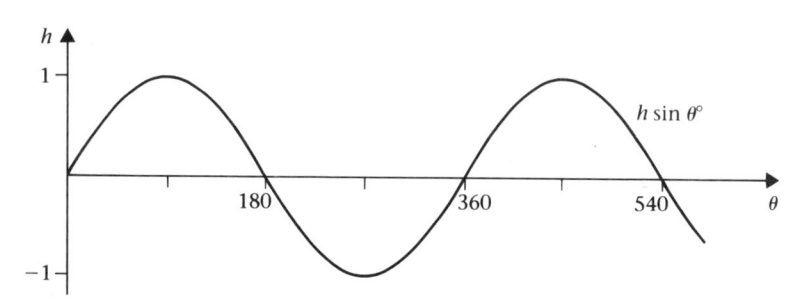

(graph of $h \sin \theta°$ from 0 to 540)

7 (a) $d = \cos 60°$, $(\theta, d) = (60, 0.50)$

   (b) (30, 0.87), (45, 0.71), (90, 0), (120, −0.50), (135, −0.71), (150, −0.87), (180, −1) (210, −0.87), (225, −0.71), (240, −0.50), (270, 0), (300, 0.5), (315, 0.71), (330, 0.87), (360, 1), (390, 0.87), (405, 0.71), (420, 0.5), (450, 0), (480, −0.5), (495, −0.71), (510, −0.87), (540, −1)

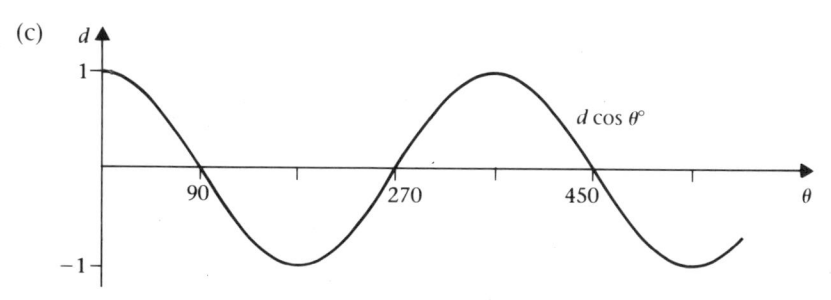

   (c) *(graph of $d \cos \theta°$ from 0 to 540)*

Transformations

TASKSHEET COMMENTARY 2

1

$y = a \sin \theta°$

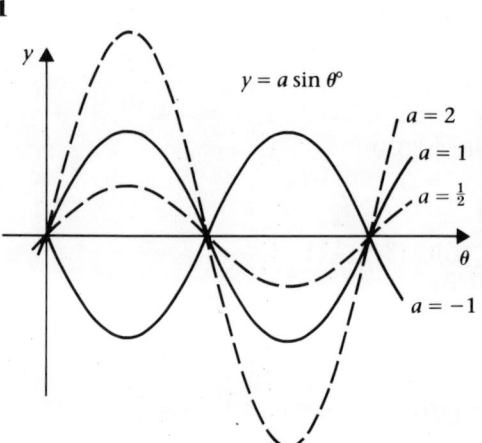

$y = \sin \theta°$ maps onto $y = a \sin \theta°$ by a stretch of factor a in the y-direction.

2 (a) A stretch in the θ-direction is needed.

$y = \sin \theta°$ maps onto $y = \sin b\theta°$ by a stretch of factor $\dfrac{1}{b}$ in the θ-direction.

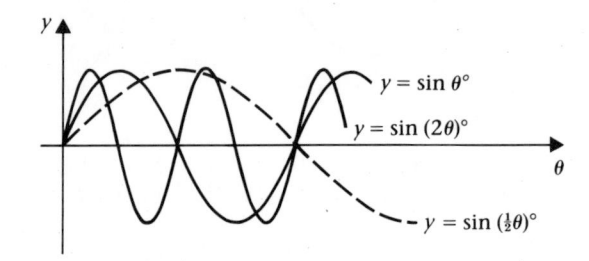

(b) The period of $\sin b\theta°$ is $\dfrac{360}{b}$.

3

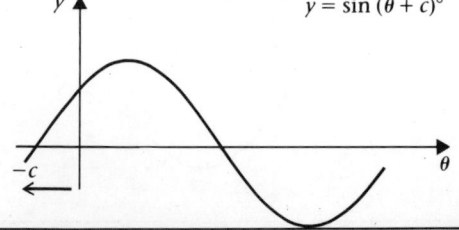

$y = \sin(\theta + c)°$

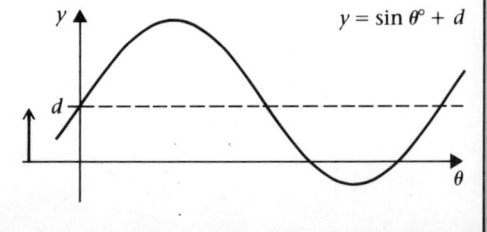

$y = \sin \theta° + d$

2 CIRCULAR FUNCTIONS

TASKSHEET COMMENTARY 2

$y = \sin(\theta + c)°$ is obtained by a translation of $-c$ in the θ-direction.
$-c$ is known as the **phase shift**.

$y = \sin \theta° + d$ is obtained by a translation of d in the y-direction.

$y = \sin(\theta + c)° + d$ is obtained by a translation of $\begin{bmatrix} -c \\ d \end{bmatrix}$.

4 (a) As in question 2, $y = \cos b\theta°$ has period $\dfrac{360}{b}$.

(b) $y = \cos \theta°$ is mapped onto $y = \cos(b\theta + c)°$ by a stretch of $\dfrac{1}{b}$ followed by a translation in the θ-direction.

Thus $y = \cos(b\theta + c)°$ has period $\dfrac{360}{b}$, and phase shift $-\dfrac{c}{b}$.

(NB. This is not surprising, since the maximum value of $\cos(b\theta + c)°$ is 1, which occurs when $b\theta + c = 0$, i.e. $\theta = -\dfrac{c}{b}$.)

5 $a = 2$, $b = 3$, $c = 30$

6 $y = \cos \theta°$ is mapped onto $y = a \cos(b\theta + c)° + d$ by stretches of a and $\dfrac{1}{b}$ in the y- and θ-directions followed by a translation $\begin{bmatrix} -c/b \\ d \end{bmatrix}$. Note that it is essential that the stretches are done before the translation.

Applications

TASKSHEET COMMENTARY 3

1 (a) $\theta = 6t$

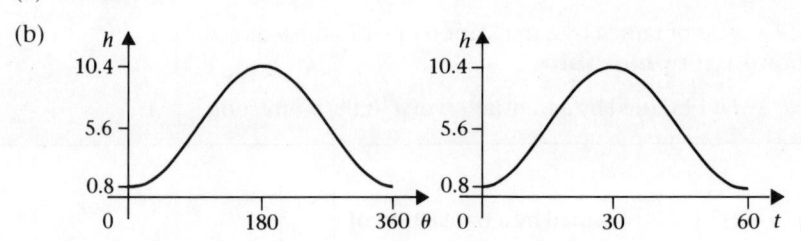

(b)

$h = 5.6 - 4.8 \cos \theta°$ $h = 5.6 - 4.8 \cos 6t°$

(c)

θ	0	30	60	90	120	150	180
h	0.8	1.4	3.2	5.6	8.0	9.8	10.4

(d)

t	0	5	10	15	20	25	30
h	0.8	1.4	3.2	5.6	8.0	9.8	10.4

2 (a) $\theta = 36t$ (b) (i) $2 \cos 36t°$ (ii) $2 + 2 \cos 36t°$

3 $l = 12 + 2.5 \cos 360t°$

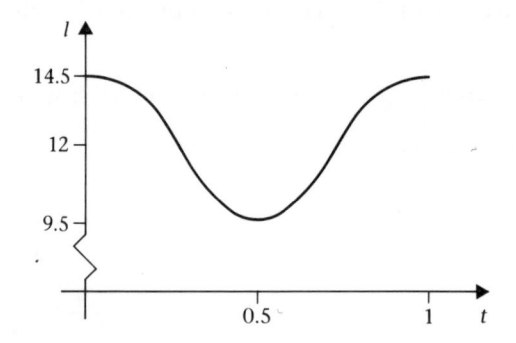

4 (a) $E = \sin \left(\dfrac{360t}{28}\right)°$ $I = \sin \left(\dfrac{360t}{33}\right)°$ (b) –

(c) Critical days occur every 11.5, 14 and 16.5 days respectively for the three cycles.

(d) All these cycles are critical about once every $7\frac{1}{4}$ years.

Inverse functions

TASKSHEET COMMENTARY 4E

1 (a) $x = 23.6$

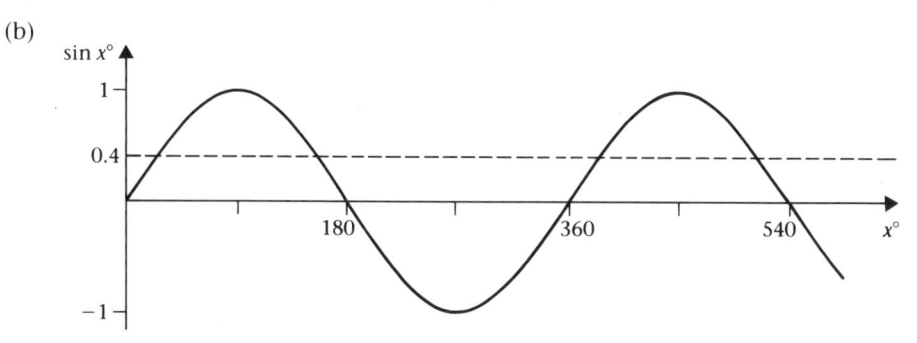

(b)

[Graph showing $\sin x°$ with dashed line at 0.4, marking 180, 360, 540 on the $x°$ axis]

From the graph the other solutions are
$180 - 23.6 = 156.4$, $360 + 23.6 = 383.6$, $540 - 23.6 = 516.4$

(c) $3600 + 23.6 = 3623.6$, $3600 + 156.4 = 3756.4$

(d) $360n + 23.6$, $360n + 156.4$

(e) Yes. For example, if $n = -1$, the solutions $23.6 - 360$ and $156.4 - 360$ are generated. The graph has period 360 in both positive and negative directions.

2 (a) $p = 30$

(b) $540 - 30$, $720 + 30$

(c) $180 \times 20 + 30 = 3600 + 30$, $180 \times 21 - 30 = 3780 - 30$

(d) $180n + 30$ if n is even, $180n - 30$ if n is odd.
Using the $(-1)^n$ notation, $x = 180n + (-1)^n 30$.

3 (a) The principal value is 44.4
The general solution is $180n + (-1)^n 44.4$

(b) The principal value is -44.4
The general solution is $180n - (-1)^n 44.4$

(c) The principal value is 45.6
Other solutions are $360 - 45.6$, $360 + 45.6$, $720 + 45.6$, $720 - 45.6$, etc.
The general solution is $360n \pm 45.6$

TASKSHEET COMMENTARY 4E

4

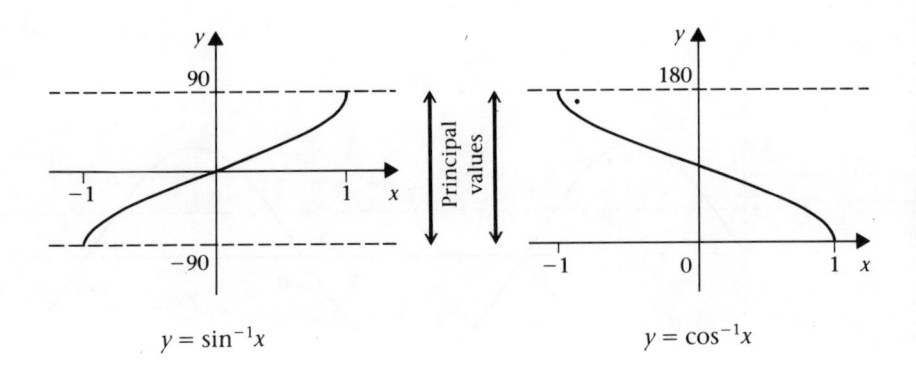

$y = \sin^{-1} x$ $\qquad\qquad$ $y = \cos^{-1} x$

$\sin^{-1} x + \cos^{-1} x = 90°$ for all x for which both functions are defined.

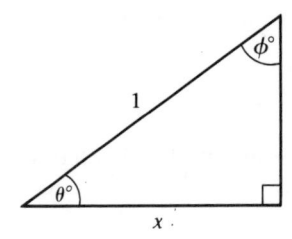

In the triangle shown,
$$\cos \theta° = x \implies \theta = \cos^{-1} x$$
Similarly $\phi = \sin^{-1} x$. But
$$\theta + \phi = 90° \implies \cos^{-1} x + \sin^{-1} x = 90°.$$

This completes the explanation in the case $0 \leq x \leq 1$.

If $-1 \leq x < 0$, then
$$\sin^{-1} x = -\sin^{-1}(-x), \quad \cos^{-1} x = 180° - \cos^{-1}(-x)$$

So
$$\sin^{-1} x + \cos^{-1} x = 180° - [\sin^{-1}(-x) + \cos^{-1}(-x)]$$
$$= 180° - 90°$$

since $(-x)$ may be taken as a side of a right-angled triangle, as in the first case.

Solving equations

TASKSHEET COMMENTARY 5S

1 $\cos x° = \frac{3}{4} = 0.75$

One solution is $x = 41.4$

The other solution between 0 and 360 is
$x = 360 - 41.4 = 318.6$

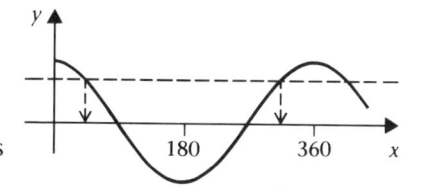

2 (a) $\cos x° = 0.56 \Rightarrow x = 55.9, 304.1$

(b) $\sin x° = -0.23 \Rightarrow x = 193.3, 346.7$

(c) $\cos x° = -0.5 \Rightarrow x = 120, 240$

3 (a) $\sin x° = 0.65 \Rightarrow x = 40.5, 139.5$

(b) $\cos x° = -0.38 \Rightarrow x = 112.3, -112.3$

(c) $\sin x° = -0.47 \Rightarrow x = -28.0, -152.0$

4 (a) $3 \sin x° = 2 \Rightarrow x = 41.8, 138.2$

(b) $5 \cos x° + 2 = 0 \Rightarrow x = 113.6, 246.4$

(c) $2 \cos x° + 5 = 0$ No solution for x

5 $5 \sin (3t + 40)° = 4$

$\Rightarrow \sin (3t + 40)° = \frac{4}{5}$

$\Rightarrow \sin x° = 0.8$, where $x = 3t + 40$

From a calculator, $x = 53.1$

$x = 53.1, 126.9, 413.1, 486.9, 773.1, 846.9$

$t = 4.37, 28.97, 124.37, 148.97, 244.37, 268.97$

6 (a) $\sin 2t° = 0.7 \Rightarrow t = 22.2, 67.8, 202.2, 247.8$

(b) $2 \cos 3t° = 1 \Rightarrow t = 20, 100, 140, 220, 260, 340$

(c) $3 \cos (0.5t + 20)° = 2 \Rightarrow t = 56.4$

2 CIRCULAR FUNCTIONS

Tan $\theta°$

TASKSHEET COMMENTARY 6S

1 (a), (b)

θ	10	20	30	40	50	60	70	80
tan $\theta°$	0.18	0.36	0.58	0.84	1.19	1.73	2.75	5.67

(c) When $\theta = 90$, tan $\theta°$ is not defined (a calculator displays 'E')

2 (a)

θ	0	15	30	45	60	75	90
$\dfrac{\sin \theta°}{\cos \theta°}$	0	0.27	0.58	1.00	1.73	3.73	undefined

(b) $\dfrac{\sin \theta°}{\cos \theta°} = \tan \theta°$. Both sides are undefined when $\theta = 90$.

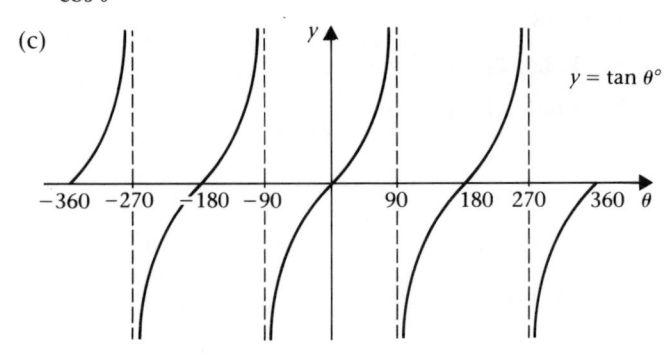

(c)

[Graph of $y = \tan \theta°$ with vertical asymptotes at odd multiples of 90°, shown from −360 to 360.]

3 $t = \tan \theta°$

(a) [Circle diagram showing angle $\theta°$ in the fourth quadrant with radius of length 1 and tangent line going downwards to point t.]

(b) [Circle diagram showing angle $\theta°$ with rotating radius projected backwards to meet the tangent line.]

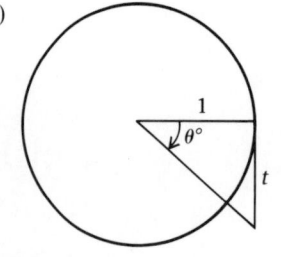

The tangent is directed downwards. t is negative.

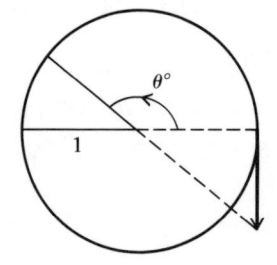

The rotating radius must be projected backwards to meet the tangent, so t is negative for $90 < \theta < 270$.

26

3 Growth functions

3.1 Exponential growth

> Estimate, using the graph, the times when the numbers of bacteria were
>
> (a) 4000, 8000 (b) 5000, 10000
>
> Estimate when the numbers of bacteria were
>
> (c) 24000 (d) 1500
>
> Describe the feature of the graph which enabled you to make these estimates.

(a) From the graph there were 4000 bacteria after approximately 2 hours and 8000 after 7 hours.

(b) There were 5000 bacteria after approximately 3.7 hours and 10000 after 8.7 hours.

(c) Notice that in each of the cases above it takes 5 hours to double the number of bacteria. If the number of bacteria doubles in any period of 5 hours, you could use this to estimate the time for 24000 bacteria. Doubling does appear to take place in all 5-hour intervals.

Since there are 12000 bacteria after 10 hours it is reasonable to suggest that there will be 24000 after 15 hours.

(d) Similarly, since there are 3000 bacteria after 0 hour, there would have been 1500 bacteria 5 hours earlier, when $t = -5$.

The number of bacteria doubles every 5 hours. It can also be seen that the number of bacteria trebles approximately every 8 hours. For example, 3000 after 0 hours, 9000 after 8 hours; 4000 after 2 hours, 12000 after 10 hours. Given any fixed time period, the number of bacteria will always increase by the same factor during that time, independent of the number at the start of the period.

3 GROWTH FUNCTIONS

3.3 Growth factors

> How can you decide if the population figures given can be closely approximated using a growth function?
>
> How can you estimate the growth factor?

The table gives the data in 10-yearly intervals. To find out whether it is suitable for modelling using a growth function, you can check to see if the 10-yearly growth factor is approximately constant.

Year	Population	Growth factor
1841	15.9	
1851	17.9	1.13
1861	20.1	1.12
1871	22.7	1.13
1881	26.0	1.15
1891	29.0	1.12
1901	32.5	1.12

The 10-yearly growth factor is roughly constant, so the data can be modelled using a growth function. To do this, it is necessary to find an estimate for the yearly growth factor.

In the sixty years from 1841 to 1901 the population grows by a factor of $\frac{32.5}{15.9} = 2.044$.

If the yearly growth factor is a, this means that

$$a^{60} = 2.044 \Rightarrow a = 2.044^{\frac{1}{60}} = 1.012$$

3.4 Logarithms

> (a) Estimate $\log_2 3$ from the graph.
>
> (b) What is (i) $\log_2 32$, (ii) $\log_{10} 1000$, (iii) $\log_3 27$?
>
> (c) How can you sketch the graph of $y = \log_2 x$?

(a) $\log_2 3 \approx 1.6$

(b) (i) Since $2^5 = 32$, $\log_2 32 = 5$

(ii) Since $10^3 = 1000$, $\log_{10} 1000 = 3$

(iii) Since $3^3 = 27$, $\log_3 27 = 3$

(c) Since $\log_2 x$ is the inverse function of 2^x, the graph of $\log_2 x$ is the reflection of the graph of 2^x in the line $y = x$.
NB. Since $\log_2 1 = 0$, the graph cuts the x-axis at $x = 1$.

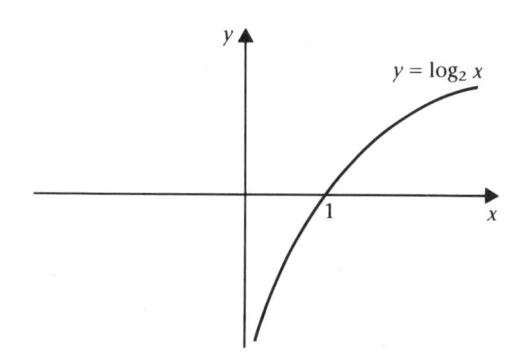

3.5 The equation $a^x = b$

(a) Initially there is 500 g of the isotope. Find an expression for the amount t years later.

(b) The half-life of the isotope is the time taken for the amount present to decrease by 50%. Use the graph to estimate this half-life.

(c) What equation must be solved to find the half-life more precisely?

(a) If 10% of the isotope decays each year then 90% will remain, so the growth factor must be 0.9. After t years, 500×0.9^t will remain.

(b) The half-life will be the time taken for the amount to drop to 250 grams. From the graph, this is approximately 6.3 years.

(c) To find the half-life more precisely, the equation

$$500 \times 0.9^t = 250$$

must be solved. This simplifies to

$$0.9^t = 0.5$$

3 GROWTH FUNCTIONS

Indices

TASKSHEET COMMENTARY 1

1 (a) (i) $2^m 2^n = \underbrace{(2 \times 2 \times \ldots \times 2)}_{m \text{ factors}} \times \underbrace{(2 \times 2 \times \ldots \times 2)}_{n \text{ factors}}$

$= \underbrace{(2 \times 2 \times \ldots \times 2)}_{(m+n) \text{ factors}} = 2^{m+n}$

(ii) A similar argument may be used, or the observation that, from (i),

$2^{m-n} 2^n = 2^m$

(b) $(2^m)^2 = 2^m \times 2^m = 2^{m+m} = 2^{2m}$

$(2^m)^3 = (2^m)^2 \times 2^m = 2^{2m+m} = 2^{3m}$

The nth term in this sequence of results is

$(2^m)^n = 2^{mn}$

2 $2^0 = 1$. Initially (when $t = 0$) the area of algae was 1 cm^2.

The result may be obtained in many ways from the rules found in part (a); for example,

$2^0 = 2^{1-1} = 2^1 \div 2^1 = 1$

3 (a) 0.5 cm, $2^{-1} = \dfrac{1}{2}$

(b) (i) $2^{-2} = \dfrac{1}{4}$ (ii) $2^{-3} = \dfrac{1}{8}$ (iii) $2^{-n} = \dfrac{1}{2^n}$

4 (a) $3^{-2} = \dfrac{1}{3^2} = \dfrac{1}{9}$ (b) $5^0 = 1$

(c) $10^{-6} = \dfrac{1}{10^6}$ (d) $a^0 = 1$ (e) $a^{-n} = \dfrac{1}{a^n}$

5 (a) $2^{-4} 2^{-3} = 2^{-7}$

(b) $2^{-4} = \dfrac{1}{16}$, $2^{-3} = \dfrac{1}{8}$

(c) $2^{-7} = \dfrac{1}{2^7} = \dfrac{1}{128} = \dfrac{1}{16} \times \dfrac{1}{8} = 2^{-4} \times 2^{-3}$

30

TASKSHEET COMMENTARY 1

(d) (i) $2^{-3} \times 2^0 = \frac{1}{8} \times 1 = \frac{1}{8} = 2^{-3} = 2^{-3+0}$

(ii) $2^{-1} \times 2 = \frac{1}{2} \times 2 = 1 = 2^0 = 2^{-1+1}$

(iii) $2^{-4} \div 2^{-3} = \frac{1}{16} \div \frac{1}{8} = \frac{1}{2} = 2^{-1} = 2^{-4-(-3)}$

(iv) $2 \div 2^{-3} = 2 \div \frac{1}{8} = 16 = 2^4 = 2^{1-(-3)}$

6 (a), (b), (d) and (e) are equal.

7 (a) $\left(2^{\frac{1}{2}}\right)^2 = 2^{\frac{1}{2} \times 2}$ (or $2^{\frac{1}{2}+\frac{1}{2}}$) $= 2^1 = 2$

(b) Since $(\sqrt{2})^2 = 2$, $2^{\frac{1}{2}}$ is defined as $\sqrt{2}$

Since $\left(2^{\frac{1}{3}}\right)^3 = 2$, $2^{\frac{1}{3}}$ is defined as $\sqrt[3]{2}$

Similarly, $2^{\frac{1}{n}} = \sqrt[n]{2}$ and $a^{\frac{1}{n}} = \sqrt[n]{a}$

8 (a) $9^{\frac{1}{2}} = \sqrt{9} = 3$

(b) $8^{\frac{1}{3}} = \sqrt[3]{8} = 2$

(c) $64^{\frac{1}{4}} = \sqrt[4]{64} = \sqrt{(\sqrt{64})} = \sqrt{8}$ (or $2\sqrt{2}$)

(d) $81^{0.5} = 9$

9 (a) $\left(4^{\frac{1}{q}}\right)^p = 4^{\frac{1}{q}} \times 4^{\frac{1}{q}} \times \ldots \times 4^{\frac{1}{q}}$ (p factors)

$= 4^{\frac{1}{q}+\frac{1}{q}+\ldots+\frac{1}{q}}$ (p terms)

$= 4^{\frac{p}{q}}$

(b) Similarly,

(i) $8^{\frac{2}{3}} = \left(8^{\frac{1}{3}}\right)^2 = 2^2 = 4$

(ii) $16^{\frac{3}{4}} = \left(16^{\frac{1}{4}}\right)^3 = 2^3 = 8$

10 Students should investigate all the laws and include cases in which an index is negative and others in which an index is not a whole number or a simple fraction.

31

Laws of indices

TASKSHEET COMMENTARY 2S

1 (a) 2^{12} (b) 2^3 (c) 2^1 (d) 2^2
     (e) 2^0 (f) 2^6 (g) 2^6 (h) 2^{16}

2 (a) y^{12} (b) b^3 (c) $c^1 = c$ (d) x^2
     (e) $y^0 = 1$ (f) a^6 (g) a^6 (h) b^{16}

3 (a) $\frac{1}{32}$ (b) $\frac{1}{5}$ (c) $\frac{1}{16}$ (d) $\frac{1}{16}$
     (e) $\frac{1}{25}$ (f) $\frac{1}{64}$ (g) 2 (h) $\frac{1}{243}$

4 (a) $x^1 = x$ (b) a^{-2} (c) $b^0 = 1$ (d) d^{-6}
     (e) x^{-10} (f) y^{-3} (g) $a^0 = 1$ (h) x^2

5 (a) 2 (b) $\frac{1}{2}$ (c) 9 (d) $\frac{1}{9}$
     (e) 8 (f) $\frac{1}{10}$ (g) 1 (h) 1000
     (i) 16 (j) 1 (k) 8 (l) $10\,000$
     (m) 27 (n) $\frac{1}{25}$ (o) 100 (p) 1000

Ka^x

TASKSHEET COMMENTARY 3

1 (a) The family of graphs is as shown. The graph of the function cuts the y-axis where $y = K$.

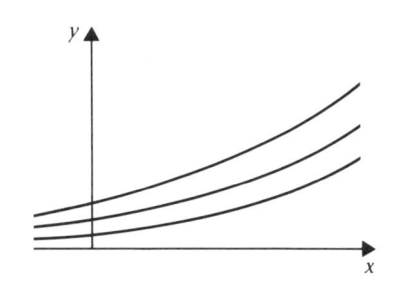

(b) If $y = K \times a^x$, $y = K$ when $x = 0$

2 (a) $y = 1$ when $t = 0$ (b) $y = K$ when $t = 0$

3 This time the graph shows exponential decay. The graph of $y = K \times (\frac{1}{2})^x$ is obtained by reflecting that of $y = K \times 2^x$ in the y-axis.

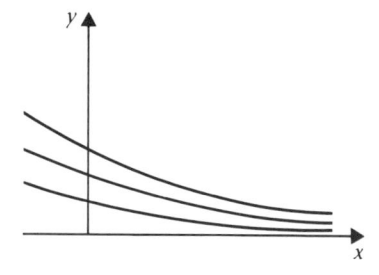

Again, K is the initial value of y.

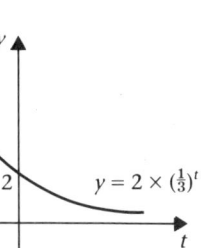

4 (a) $y = 5 \times 3^t$ (b) $y = 2 \times (\frac{1}{3})^t$ (c) $y = \frac{1}{2} \times 5^t$ (d) $y = 2 \times (\frac{1}{5})^t$

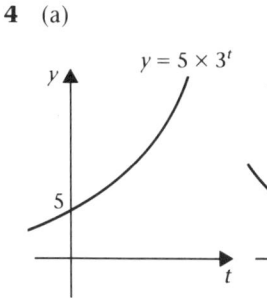

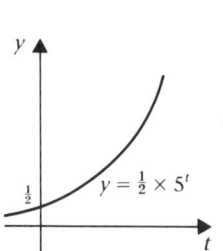

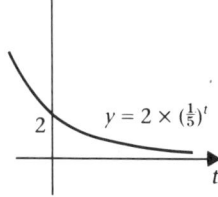

TASKSHEET COMMENTARY 3

5 (a) $K = 1.5 \times 10^6$, i.e. the initial value.

(b) In 1700, $P = 6.1 \times 10^6 = Ka^{634}$, since $t = 634$

i.e. $6.1 \times 10^6 = 1.5 \times 10^6 \, a^{634}$

$\Rightarrow a^{634} = \dfrac{6.1}{1.5}$

$\Rightarrow a = \left(\dfrac{6.1}{1.5}\right)^{\frac{1}{634}} = 1.00222$

(c) In 1990, $t = 924$

$P = Ka^{924} = 1.5 \times 10^6 \times 1.00222^{924}$

$= 11.59 \times 10^6$

(d) The annual growth factor since 1700 has been considerably greater than 1.00222, since the population of the United Kingdom is approximately 55 million.

6 If $V = Ka^t$,

$V = 15$ when $t = 0 \Rightarrow K = 15$

Growth factor $a = \left(\dfrac{6}{15}\right)^{\frac{1}{12}} \approx 0.93$

$V = 15 \times \left(\dfrac{6}{15}\right)^{\frac{t}{12}} \approx 15 \times 0.93^t$

Properties of logs

TASKSHEET COMMENTARY 4

1. (a) $2^6 = 64 \Rightarrow \log_2 64 = 6$
   (b) $2^{-3} = \frac{1}{8} \Rightarrow \log_2 \frac{1}{8} = -3$
   (c) $2^1 = 2 \Rightarrow \log_2 2 = 1$
   (d) $2^{\frac{1}{2}} = \sqrt{2} \Rightarrow \log_2 \sqrt{2} = \frac{1}{2}$

2. (a) $3^2 = 9 \Rightarrow \log_3 9 = 2$
   (b) 3
   (c) -2
   (d) 0
   (e) -3
   (f) $\frac{1}{4}$, since $\sqrt[4]{3} = 3^{\frac{1}{4}}$
   (g) $\frac{1}{2}$, since $2 = 4^{\frac{1}{2}}$
   (h) 1
   (i) cannot be found since 3^x is always positive

3. (a) $a^0 = 1 \Rightarrow \log_a 1 = 0$
   (b) $a^1 = a \Rightarrow \log_a a = 1$
   (c) $a^{-1} = \frac{1}{a} \Rightarrow \log_a \frac{1}{a} = -1$
   (d) $\log_a a^2 = 2$

4. (a) $\log_{10} 10^{3.7} = 3.7$
   (b) $10^{\log_{10} 3.7} = 3.7$

   This is to be expected, since $\log_{10} x$ and 10^x are inverse functions. In each case 3.7 is operated on by the function and its inverse and so is its own image.

5. (a) (i) $\log_2 8 = 3$ (ii) $\log_2 16 = 4$ (iii) $\log_2 128 = 7$
   (b) $8 \times 16 = 128$ becomes $2^3 \times 2^4 = 2^7$ so $a = 3$, $b = 4$, $c = 7$ and $a + b = c$.
   (c) Since $a = \log_2 8$, $b = \log_2 16$, $c = \log_2 128$ it follows that
   $\log_2 8 + \log_2 16 = \log_2 128$.

TASKSHEET COMMENTARY 4

6 As with question 5, 2 + 3 = 5, but 2 = $\log_3 9$, 3 = $\log_3 27$ and
5 = $\log_3 243 = \log_3 (9 \times 27)$ so $\log_3 9 + \log_3 27 = \log_3 (9 \times 27)$.

7 (b) (i) $\log_{10} 3 = 0.4771$ (ii) $\log_{10} 5 = 0.6990$
(c) $\log_{10} 15 = \log_{10} (3 \times 5) = \log_{10} 3 + \log_{10} 5 = 0.4771 + 0.6990 = 1.1761$

8 $\log_{10} 9 = 0.954$, $\log_{10} 8 = 0.903$, $\log_{10} 72 = 1.857 = 0.954 + 0.903$

9 $\log_a m + \log_a \dfrac{l}{m} = \log_a m \times \dfrac{l}{m} = \log_a l$

$\Rightarrow \log_a l - \log_a m = \log_a \dfrac{l}{m}$

10 $\log_{10} \dfrac{1}{2} = \log_{10} 1 - \log_{10} 2 = 0 - 0.3010 = -0.3010$

$\log_{10} 1.5 = \log_{10} \dfrac{3}{2} = \log_{10} 3 - \log_{10} 2 = 0.4771 - 0.3010 = 0.1761$

$\log_{10} 2.5 = \log_{10} \dfrac{10}{4} = \log_{10} \dfrac{10}{2 \times 2} = \log_{10} 10 - \log_{10} 2 - \log_{10} 2$
$= 1 - 2 \times 0.3010 = 0.398$

$\log_{10} 4 = \log_{10} (2 \times 2) = \log_{10} 2 + \log_{10} 2 = 2 \times 0.3010 = 0.6020$

$\log_{10} 5 = \log_{10} \dfrac{10}{2} = \log_{10} 10 - \log_{10} 2 = 1 - 0.3010 = 0.6990$

$\log_{10} 6 = \log_{10} (2 \times 3) = \log_{10} 2 + \log_{10} 3 = 0.3010 + 0.4771 = 0.7781$

$\log_{10} 8 = \log_{10} (2 \times 2 \times 2) = 3 \log_{10} 2 = 3 \times 0.3010 = 0.9030$

$\log_{10} 9 = \log_{10} (3 \times 3) = 2 \log_{10} 3 = 2 \times 0.4771 = 0.9542$

$a^x = b$

TASKSHEET COMMENTARY 5

1 (a) (i) $\log_{10} 49 = 1.6902$, $\log_{10} 7 = 0.8451$, $\log_{10} 49 = 2 \log_{10} 7$

(ii) $\log_{10} 64 = 1.806$, $\log_{10} 2 = 0.301$, $\log_{10} 64 = 6 \times \log_{10} 2$

(iii) $\log_{10} 125 = 2.0969$, $\log_{10} 5 = 0.6990$, $\log_{10} 125 = 3 \times \log_{10} 5$

(b) The results above suggest that $\log_{10} m^p = p \log_{10} m$.

2 (a) $\log_a m^2 = \log_a (m \times m) = \log_a m + \log_a m = 2 \log_a m$

(b) For any positive index,

$$\log_a m^p = \log_a \underbrace{m \times m \times \ldots \times m}_{p \text{ factors}}$$

$$= \underbrace{\log_a m + \log_a m + \ldots + \log_a m}_{p \text{ terms}}$$

$$= p \log_a m$$

3 $\log 2^x = x \log 2$

If $2^x = 7$, then $\log 2^x = \log 7$

$\Rightarrow x \log 2 = 7$

$\Rightarrow x = \dfrac{\log 7}{\log 2} = 2.807$

TASKSHEET COMMENTARY 5

4 (a) 1% interest represents a monthly growth factor of 1.01.
After m months the amount in the account will be 1000×1.01^m
There will be £2000 in the account when $1000 \times 1.01^m = 2000$.
Dividing by 1000,
$$1.01^m = 2$$

(b) $$1.01^m = 2$$
$$\log(1.01^m) = \log 2$$
$$m \log 1.01 = \log 2$$
$$m = \frac{\log 2}{\log 1.01} = 69.66$$

The amount will be a little over £2000 (£2006.76) after 70 months.

5 The equation to find the half-life of bismuth-210 is
$$10 \times (0.87)^t = 5$$
or
$$(0.87)^t = 0.5$$
$$\log(0.87^t) = \log 0.5$$
$$t \log(0.87) = \log 0.5$$
$$t = \frac{\log 0.5}{\log 0.87} = 4.98$$

The half-life is 5.0 days.

4 Radians

4.1 Rates of change

> (a) What is the equation of the gradient graph?
>
> (b) What is the connection between k and $\dfrac{\sin \theta°}{\theta}$?
>
> (c) How can you obtain a good estimate of k?

(a) The gradient graph appears to be a cosine curve, with amplitude k. It has equation

$$\frac{dy}{d\theta} = k \cos \theta°$$

(b) k is the gradient of $y = \sin \theta°$ at 0. Thus, from the diagram and argument above

$$k \approx \frac{\sin \theta°}{\theta}$$

(c) You can obtain a sequence of values which approaches k by taking smaller and smaller values of θ. This is explored further in tasksheet 1.

4.3 Area and arc lengths

> (a) What is the length of arc AB?
> (b) What angle θ gives an arc length of 1 unit?
> (c) Suggest a value for the area of sector AOB.

(a) Since the circumference of a full circle is 2π units and the arc is $\dfrac{\theta}{2\pi}$ of the full circle, the length of arc AB is $\dfrac{\theta}{2\pi} \times 2\pi = \theta$.

(b) If the arc length is 1, the angle θ at the centre is also 1. This gives rise to the alternative definition of a radian as that angle which is subtended at the centre by an arc of length 1 in a circle of radius 1.

(c) $\frac{1}{2}\theta$. In general, the area of a sector of a circle of radius r is $\frac{1}{2}r^2\theta$.

4.5 Applications

> (a) What is the period of the function which describes the height of the tide?
> (b) If the height of the tide is h metres at time t hours, then explain why $h = 2.5 \sin \frac{1}{6}\pi t + 5$

(a) 12 hours

(b) Since the period of $y = \sin \omega t$ is $\dfrac{2\pi}{\omega}$,

$$\dfrac{2\pi}{\omega} = 12 \Rightarrow \omega = \frac{1}{6}\pi$$

The amplitude is 2.5 and the mean height is 5 and so
$$h = 2.5 \sin \tfrac{1}{6}\pi t + 5$$

Differentiation of sin $\theta°$

TASKSHEET COMMENTARY 1

1 (a) (i)

$\theta°$	10	5	2	1	0.1
$\dfrac{\sin \theta°}{\theta}$	0.01736	0.01734	0.01745	0.01745	0.01745

(ii) The sequence tends to a value 0.01745 rounded to five decimal places.

2 (a)

θ	10	5	2	1	0.1
(i) BC	0.1736	0.0872	0.0349	0.01745	0.001745
(ii) arc BA	0.1745	0.0873	0.0349	0.01745	0.001745

(b) The lengths of BC and arc BA become closer in value as θ gets smaller.

3 Length BC $= \sin \theta°$

Arc BA $= \dfrac{\theta}{360} \times 2\pi = \dfrac{\pi\theta}{180}$

Since the lengths are approximately equal for small values of θ,

$$\sin \theta° \approx \dfrac{\pi\theta}{180} \Rightarrow \dfrac{\sin \theta°}{\theta} \approx \dfrac{\pi}{180}$$

4 $\dfrac{\pi}{180} = 0.01745$, rounded to five decimal places.

5 The gradient of $\sin \theta°$ at the origin is $\dfrac{\pi}{180}$.

6 $y = \sin \theta° \Rightarrow \dfrac{dy}{d\theta} = \dfrac{\pi}{180} \cos \theta°$

Radian measure

TASKSHEET COMMENTARY 2

1. (a) 1 radian = 57.296° (to three decimal places)
   (b) 1 degree = 0.01745 radians (to five decimal places)

2. (a) 0.55 (b) 2.3 (c) $\frac{1}{2}\pi$ (d) $\frac{1}{3}\pi$ (e) 5.5 (f) 4.03

3. –

4. (a) 0.841 (b) 0.017 (c) 0.996 (d) 0.284 (e) 1.000 (f) 0.014

5. (a) $\sin 30° = \sin (\frac{1}{6}\pi)^c$

   (b)

Radians	π	$\frac{1}{2}\pi$	$\frac{1}{3}\pi$	$\frac{1}{4}\pi$	$\frac{1}{6}\pi$	$\frac{3}{2}\pi$	2π
Degrees	180	90	60	45	30	270	360

   (c) To convert $\theta°$ to radians, multiply by $\dfrac{\pi}{180}$.

   (d) To convert θ^c to degrees, multiply by $\dfrac{180}{\pi}$.

6. (a) 1.047 (b) 0.866 (c) 0.018

7. Sin 60° = 0.866. In radian mode, this will probably give an error on a calculator. In fact, $\sin 60^c = -0.305$.

8. Plotting a graph of $y = \sin x$ with x in radians, using simple numerical values rather than multiples of π, helps to emphasise that multiples of π are not the only way to express angles measured in radians.

Further derivatives

TASKSHEET COMMENTARY 3

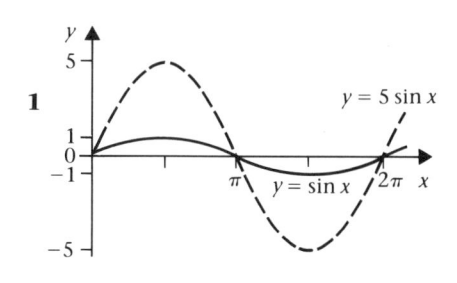

1 (a) $y = \sin x$ is mapped onto $y = 5 \sin x$ by a stretch of factor 5 in the y-direction.

(b) The gradient is multiplied by a factor of 5.

(c) $y = 5 \sin x \Rightarrow \dfrac{dy}{dx} = 5 \cos x$

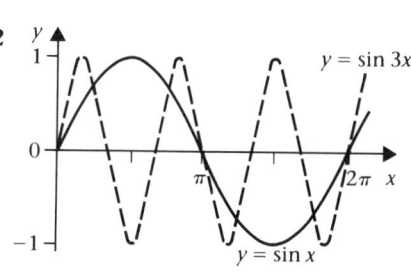

2 (a) $y = \sin x$ is mapped onto $y = \sin 3x$ by a stretch of factor $\frac{1}{3}$ in the x-direction.

(b) The gradient is multiplied by a factor of 3.

(c) $y = \sin 3x \Rightarrow \dfrac{dy}{dx} = 3 \cos 3x$

3 $y = 5 \sin 3x$ is obtained from $y = \sin x$ by applying both stretches from the previous questions.

$$y = 5 \sin 3x \Rightarrow \dfrac{dy}{dx} = 15 \cos 3x$$

4 (a) $-2 \sin 2x$ (b) $20 \cos 2x$ (c) $0.5 \cos 0.5x$

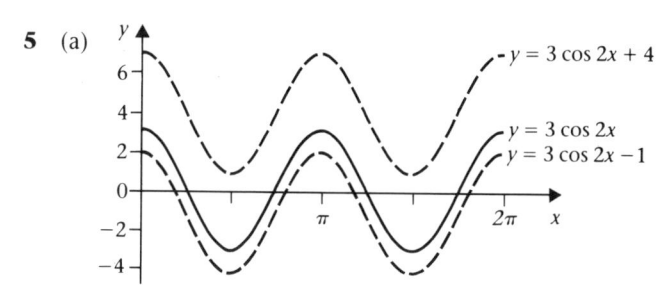

5 (a) [graph showing $y = 3 \cos 2x + 4$, $y = 3 \cos 2x$, $y = 3 \cos 2x - 1$]

(b) These curves are mapped onto each other by translations in the y-direction.

The derivatives are all the same:

$$\dfrac{dy}{dx} = -6 \sin 2x$$

6 (a) (i) $y = a \sin x \Rightarrow \dfrac{dy}{dx} = a \cos x$

(ii) $y = \sin bx \Rightarrow \dfrac{dy}{dx} = b \cos bx$

(iii) $y = a \sin bx \Rightarrow \dfrac{dy}{dx} = ab \cos bx$

(b) $y = a \cos bx \Rightarrow \dfrac{dy}{dx} = -ab \sin bx$

Derivative of sin² x

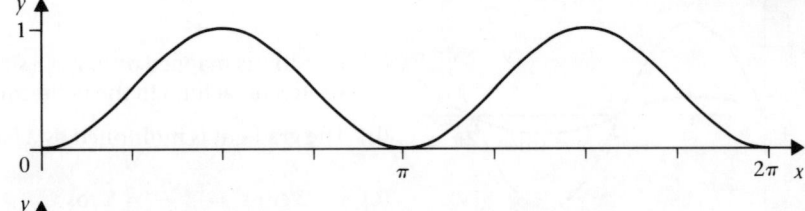

TASKSHEET COMMENTARY 4E

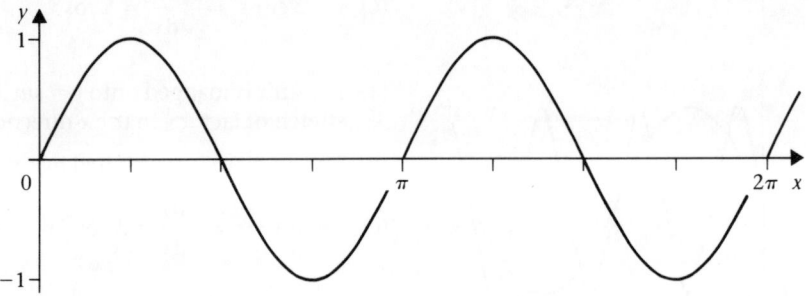

1 (a) [graph of y against x, showing sin²x from 0 to 2π]

(b) [graph of y against x, showing sin 2x from 0 to 2π]

The derivative of $\sin^2 x$ is $\sin 2x$.

2 (a) $\sin^2 x = \frac{1}{2} - \frac{1}{2}\cos 2x \Rightarrow a = -\frac{1}{2}, b = 2, c = \frac{1}{2}$

(b) If $y = \frac{1}{2} - \frac{1}{2}\cos 2x$, $\dfrac{dy}{dx} = -\frac{1}{2}(-2\sin 2x) = \sin 2x$.

3 $\dfrac{d}{dx}(\cos^2 x) = -\sin 2x$, using a method similar to that used in question 1.

$$\cos^2 x = \tfrac{1}{2} + \tfrac{1}{2}\cos 2x \Rightarrow \frac{d}{dx}(\cos^2 x) = \tfrac{1}{2}(-\tfrac{1}{2}\sin 2x) = -\sin 2x$$

4 (a) $\dfrac{d}{dx}(\sin^2 x + \cos^2 x) = \sin 2x - \sin 2x = 0$

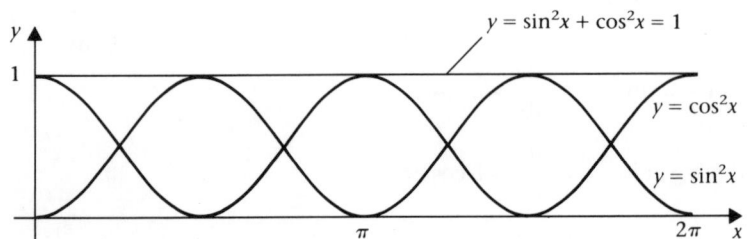

(b) [graph showing $y = \sin^2 x + \cos^2 x = 1$, $y = \cos^2 x$, $y = \sin^2 x$ from 0 to 2π]

(c) $\sin^2 x + \cos^2 x = 1$, a constant whose derivative is zero.

5 e

5.1 e^x

> (a) Is the rate at which the colony is growing when it contains 1000 bacteria less than, equal to or greater than 1000 per hour?
>
> (b) When this population is 2000, is the rate of growth less than, equal to or greater than 2000 per hour?
>
> (c) What is meant by the rate of growth and how can you calculate it?
>
> (d) Will the rate of growth and the size of the population always be related as in parts (a) and (b)?

(a) The rate of growth is less than 1000 per hour. Although in the first hour the population grows by 1000, its rate of growth is less than this to begin with and greater than 1000 as the graph becomes steeper.

(b) As in (a) the rate of growth is less than 2000 per hour.

(c) The rate of growth is the gradient of the graph when you zoom in. If you know the equation of the graph then you can find the derivative. If you do not know the equation of the graph, you must measure the gradient of the tangent line.

(d) The rate of growth is always less than the size of the population. Since in every hour the population doubles in size, the actual growth over each hour will equal the size of the population at the beginning of that hour. However, the rate of growth at the start of any period will always be less than the average rate over that period.

5.2 e^{ax}

(a) How are 2^{2t} and 4^t related? Why?

(b) Can you find alternative expressions of the form 2^{at} to represent (i) 8^t (ii) 5^t?

(c) Is it **always** possible to express b^t in the form 2^{at}?

(a) $2^{2t} = 4^t$ since $2^{2t} = (2^2)^t = 4^t$

(b) (i) $8^t = (2^3)^t = 2^{3t}$

(ii) $5^t = 2^{at} \Rightarrow 5 = 2^a$, i.e. $a = \dfrac{\log 5}{\log 2} = 2.32$

Therefore $5^t = 2^{2.32t}$

(c) When $b > 0$, it is clear that you can always write b^t in the form 2^{at}, since you can always solve the equation $b = 2^a$. As has been noted already, when $b < 0$ the meaning of b^t is not defined for some values of t (for example, $t = \tfrac{1}{2}$).

Rates of growth

TASKSHEET COMMENTARY 1

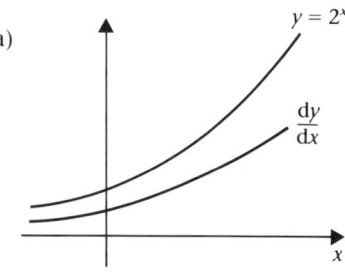

1 (a) [graph of $y = 2^x$ and $\frac{dy}{dx}$]

(b) $\frac{dy}{dx} \approx 0.7 \times 2^x$

2 (a) [graph of $y = 3^x$ and $\frac{dy}{dx}$]

$\frac{dy}{dx} \approx 1.1 \times 3^x$

(b) [graph of $y = 1.5^x$ and $\frac{dy}{dx}$]

$\frac{dy}{dx} \approx 0.4 \times 1.5^x$

(c) [graph of $y = 10^x$ and $\frac{dy}{dx}$]

$\frac{dy}{dx} \approx 2.3 \times 10^x$

3 (a)

a	$\frac{d}{dx}(a^x)$
1.5	0.4×1.5^x
2	0.7×1.2^x
3	1.1×3^x
10	2.3×10^x

(b) It is clear that a will lie somewhere between 2 and 3, probably around 2.7.

4 (a) $y = 2e^x$ is the graph of $y = e^x$ scaled by a factor of 2 in the y-direction.

(b) Since $\frac{d}{dx}(e^x) = e^x$, $\frac{d}{dx}(2e^x) = 2e^x$.

(c) If the scaling is by a factor of k, the gradient function will also be scaled by k:

$$\frac{d}{dx}(ke^x) = ke^x$$

Rates of growth

TASKSHEET COMMENTARY 1E

1 (a) [graph of $y = 2^x$ with $\frac{dy}{dx}$] (b) $\frac{dy}{dx} \approx 0.7 \times 2^x$

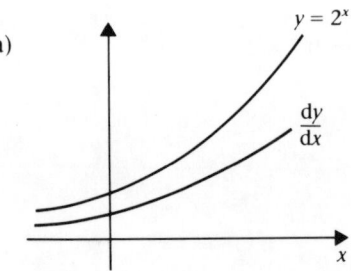

2 (a) [graph of $y = 3^x$ with $\frac{dy}{dx}$] (b) [graph of $y = 1.5^x$ with $\frac{dy}{dx}$] (c) [graph of $y = 10^x$ with $\frac{dy}{dx}$]

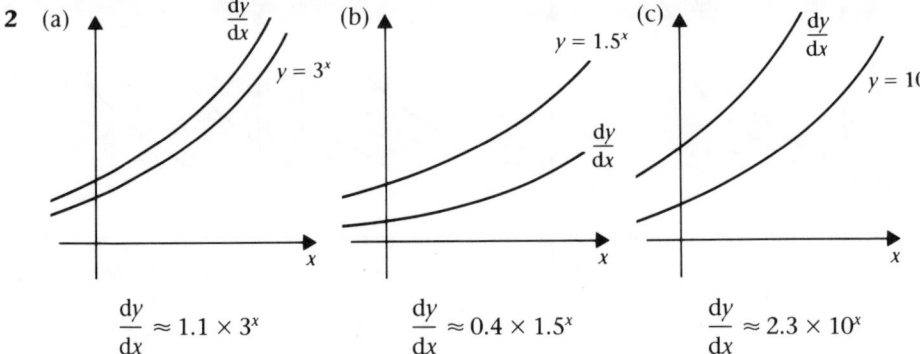

$\frac{dy}{dx} \approx 1.1 \times 3^x$ $\frac{dy}{dx} \approx 0.4 \times 1.5^x$ $\frac{dy}{dx} \approx 2.3 \times 10^x$

3

a	$\frac{d}{dx}(a^x)$
1.5	0.4×1.5^x
2	0.7×2^x
3	1.1×3^x
10	2.3×10^x

The value of a will lie somewhere between 2 and 3, probably around 2.7.

4 Since multiplication by k represents a scaling by a factor of k in the y-direction, the gradient will also be multiplied by k:

$$\frac{d}{dx}(ke^x) = ke^x$$

TASKSHEET COMMENTARY 1E

5 (a) The x-coordinate of Q is $x + 0.001$

$\Rightarrow$ the y-coordinate of Q is $2^{x + 0.001} = 2^{0.001} \times 2^x$.

(b) The gradient is

$$\frac{\text{change in } y}{\text{change in } x} = \frac{2^{0.001} \times 2^x - 2^x}{0.001}$$

(c) $\dfrac{2^{0.001} \times 2^x - 2^x}{0.001} = \dfrac{(2^{0.001} - 1)2^x}{0.001} = \dfrac{2^{0.001} - 1}{0.001} \times 2^x \approx 0.693 \times 2^x$

(d) You could zoom in even more so that the difference in x between P and Q is even less than 0.001. For example,

$$\frac{2^{0.00001} - 1}{0.00001} \times 2^x \approx 0.6931 \times 2^x$$

6 (a) $\dfrac{5^{0.00001} - 1}{0.00001} \times 5^x \approx 1.6094 \times 5^x$

(b) $y = a^x \Rightarrow \dfrac{dy}{dx} \approx \dfrac{a^{0.001} - 1}{0.001} \times a^x$

but $y = e^x \Rightarrow \dfrac{dy}{dx} = e^x$

Therefore

$$\frac{e^{0.001} - 1}{0.001} \approx 1$$

49

e^{ax}

TASKSHEET COMMENTARY 2

1 (i) $a = 1.46$ (ii) $a = 1.77$ (iii) $a = 0.63$

2 (a)

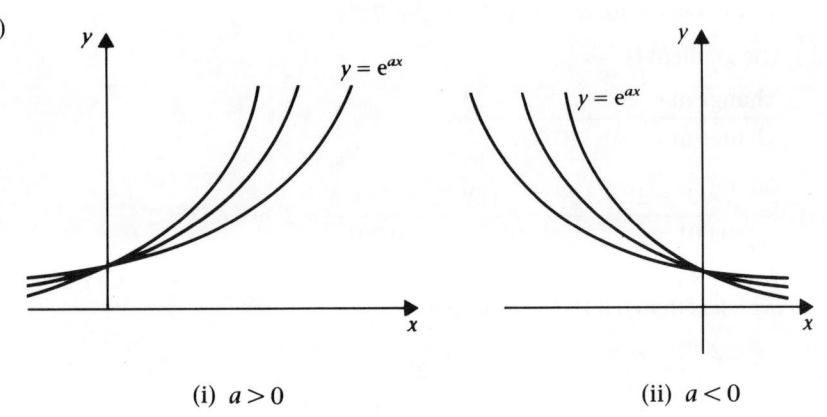

(i) $a > 0$ (ii) $a < 0$

If $a = 0$, $y = e^0 = 1$. The graph is a straight line.

(b) If $a > b > 0$, then e^{ax} lies above e^{bx} for positive values of x and below e^{bx} for negative values.

3 (a) 1.61 (b) 2.08 (c) 0.69

4 (a) $\dfrac{d}{dx}(e^{0.69x}) = \dfrac{d}{dx}(2^x) = 0.69 \times 2^x = 0.69 \times e^{0.69x}$

(b) As in (a), $\dfrac{d}{dx}(e^{5x}) = 5 e^{5x}$

5 (a) e^{2q} (b) q

(c) It multiplies the gradient of the graph by 2.

(d) $2g$, from part (c)

(e) The gradient is e^x, which at R is e^{2q}. Therefore the gradient at Q is $2 \times e^{2q}$.

(f) $\dfrac{d}{dx}(e^{2x}) = 2 \times e^{2x}$

Properties of ln x

TASKSHEET COMMENTARY 3

1 (a) (i) $x = \ln 1 \Rightarrow e^x = 1 \Rightarrow x = 0$

(ii) $\ln e = \ln e^1 = 1$

(iii) $\ln e^2 = 2$

(iv) $\ln \dfrac{1}{e} = \ln e^{-1} = -1$

(v) $\ln \dfrac{1}{e^5} = \ln e^{-5} = -5$

(vi) $\ln(-1) = x \Rightarrow e^x = -1$, which has no solution. i.e. $\ln(-1)$ is not defined.

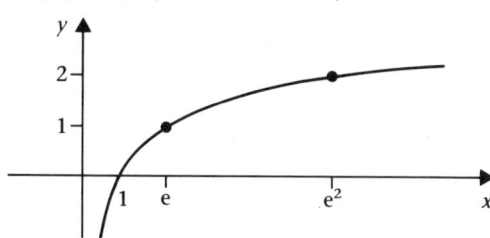

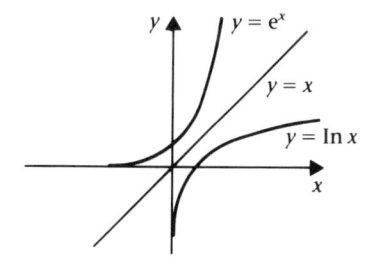

2 Since $y = e^x$ and $y = \ln x$ are inverse functions, one is the reflection of the other in the line $y = x$.

3 (a) $b = e^a$ (b) (b, a)

(c) From the triangle with sides α, β, the gradient of $y = e^x$ at P is $\dfrac{\beta}{\alpha}$. From the triangle with sides α, β, the gradient of $y = \ln x$ at Q is $\dfrac{\alpha}{\beta}$.

(d) Using a result in (c), the gradient is $\dfrac{1}{e^a}$.

(e) The gradient of $y = \ln x$ at Q is $\dfrac{1}{e^a} = \dfrac{1}{b}$. The x-coordinate of Q is b and so

$$\dfrac{d}{dx}(\ln x) = \dfrac{1}{x}.$$

6 Transformations

6.1 Graph sketching

> Use a graph plotter to sketch the equation before and after the transformations.
>
> What single geometric transformation is produced by replacing x with $-x$ and y with $-y$?

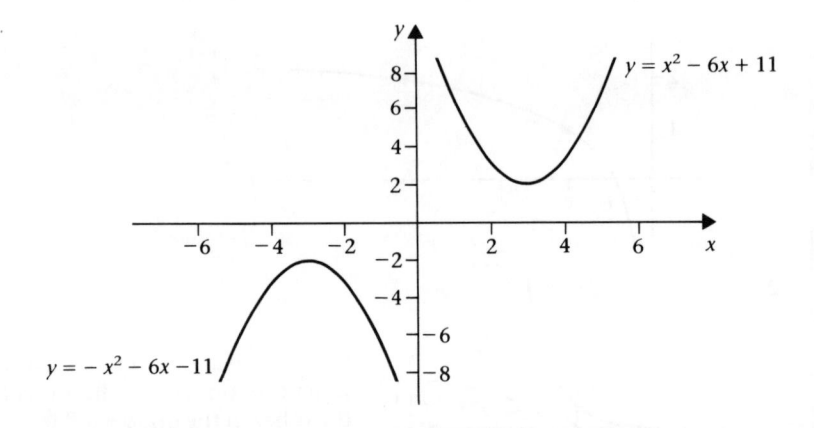

Replacing x with $-x$ and y with $-y$ rotates the graph 180° about the origin. This ties in with the idea of an odd function which you met earlier in the unit.

For a function $y = f(x)$, replacing x with $-x$ and y with $-y$ gives $-y = f(-x)$, i.e. $y = -f(-x)$. If $f(x) = -f(-x)$, then the function has 180° rotational symmetry and is therefore an odd function.

6.2 Stretching a circle

(a) What is the equation of a circle, radius r, centre $(0, 0)$?

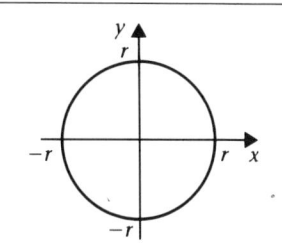

(b) What is the equation of the ellipse shown in the diagram?

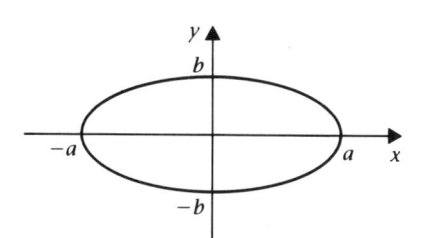

(c) What is the equation of a circle radius r, centre (p, q)?

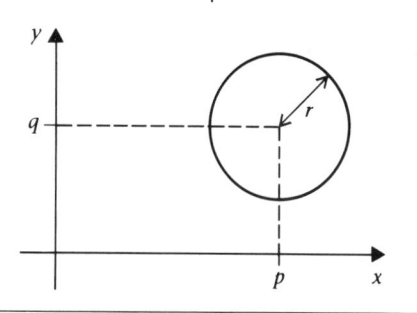

(a) If you transform the unit circle, $x^2 + y^2 = 1$, with a stretch factor r from the y-axis and a stretch factor r from the x-axis then you obtain the circle with radius r shown in the diagram.

Replacing x with $\dfrac{x}{r}$ and y with $\dfrac{y}{r}$ in the equation $x^2 + y^2 = 1$ gives

$$\frac{x^2}{r^2} + \frac{y^2}{r^2} = 1 \Rightarrow x^2 + y^2 = r^2$$

(b) Transforming the unit circle, $x^2 + y^2 = 1$, by a stretch factor a along the x-axis and by a stretch factor b along the y-axis gives

$$\frac{x^2}{a^2} + \frac{y^2}{b^2} = 1$$

(c) Giving the circle in (a) a translation $\begin{bmatrix} p \\ q \end{bmatrix}$ by replacing x with $x - p$ and y with $y - q$, you obtain

$$(x - p)^2 + (y - q)^2 = r^2$$

Transforming equations

TASKSHEET COMMENTARY 1

1. (a) Replacing x with $x + k$ results in the translation $\begin{bmatrix} -k \\ 0 \end{bmatrix}$.

   (i) $y = \frac{1}{2}(x + k)^2 - 4(x + k)$ (ii) $y \sin(x + k)$
   (iii) $y = e^{x + k}$ (iv) $y = \pm \sqrt{(1 - (x + k)^2)}$

   (b) Replacing y with $y + k$ results in the translation $\begin{bmatrix} 0 \\ -k \end{bmatrix}$.

   (i) $y + k = \frac{1}{2}x^2 - 4x \Rightarrow y = \frac{1}{2}x^2 - 4x - k$
   (ii) $y + k = \sin x \Rightarrow y = \sin x - k$
   (iii) $y + k = e^x \Rightarrow y = e^x - k$
   (iv) $y + k = \pm\sqrt{(1 - x^2)} \Rightarrow y = -k \pm \sqrt{(1 - x^2)}$

   (c) Replacing x with kx results in a one-way stretch from the y-axis with scale factor $\frac{1}{k}$.

   (i) $y = \frac{1}{2}(kx)^2 - 4kx$ (ii) $y = \sin kx$
   (iii) $y = e^{kx}$ (iv) $y = \pm\sqrt{(1 - (kx)^2)}$

   (d) Replacing y with ky results in a one-way stretch from the x-axis with scale factor $\frac{1}{k}$.

   $ky = \dfrac{x^2}{2} + 4x \Rightarrow y = \dfrac{x^2}{2k} - \dfrac{4x}{k}$

   $ky = \sin x \Rightarrow y = \dfrac{\sin x}{k}$

   $ky = e^x \Rightarrow y = \dfrac{e^x}{k}$

   $ky = \pm\sqrt{(1 - x^2)} \Rightarrow y = \pm\dfrac{\sqrt{(1 - x^2)}}{k}$

2. (a) Replacing x with $-x$ reflects the graph in the y-axis.
   (b) Replacing y with $-y$ reflects the graph in the x-axis.
   (c) Interchanging x and y reflects the graph in the line $y = x$.